A Deed without a Name

A Deed without a Name

The Witch in Society and History

Andrew Sanders

BERG
Oxford • Washington, D.C.

First published in 1995 by
Berg Publishers Limited
Editorial offices:
150 Cowley Road, Oxford, OX4 1JJ, UK
13950 Park Center Road, Herndon, VA 22071, USA

© Andrew Sanders 1995

All rights reserved.
No part of this publication may be reproduced in any form
or by any means without the written permission of
Berg Publishers Limited.

Library of Congress Cataloging-in-Publication Data

A catalogue record for this book is available from the Library of Congress.

British Library Cataloguing-in-Publication Data

A catalogue record for this book is available from the British Library.

ISBN 1 85973 048 5 (Cloth)
1 85973 053 1 (Paper)

Printed in the United Kingdom by WBC Bookbinders, Bridgend,
Mid Glamorgan.

For Judy and Dave
who really deserve a book each

Macbeth: How now, you secret, black, and midnight hags?
What is't you do?
Three Witches: A deed without a name.

Macbeth, Act 4, Scene 1

Contents

	Preface	ix
1	**Power and the Witch**	1
	The Cases	2
	Conflict Theory	4
2	**The Witch and Society**	10
	Witch Beliefs and the Structure of Society	24
3	**The Idea of the Witch**	30
	The Witches' Society	41
	Symbolic Anthropologists and the Witch-Image	49
4	**Detecting the Witch**	54
	Anti-Witchcraft Associations	67
	Ordeals	71
5	**Witch Suspects: Suspicions Based upon the Personal Qualities of Suspects**	73
	Witchfinders and Curers	74
	Antisocial Persons	83
	Functionalist Explanations of Witchcraft	90
6	**Witch Suspects: Suspicions Rooted in the Structural Position of the Accused**	93
	Strangers and Foreigners	93
	Ambiguous and Marginal Social Positions	94
	Personal Relations of Conflict	101
	Political Competitors	110
	Self-Confessed Witches	110
7	**Labelling the Witch**	113
	A Ndembu Case Study	122

Contents

8	**Witchcraft, Power and Wealth**	130
	Witchcraft against Kings and Powerful Personages	130
	Witch-Hunts in Indigenous Societies	134
	Accusing Political Competitors – the Efik of Old Calabar	138
9	**European Witchcraft: *Maleficium* and Demonology**	145
	Classical Europe, the Dark Ages, and the Early Middle Ages	145
	The 'Classic' European Witch	149
	Heresy and Devil-Worship	151
	Political Trials Involving Beliefs that became Incorporated into Late Medieval Witchcraft	157
	Maleficium into Diabolic Witchcraft	165
10	**'The Great Witch-Craze'**	169
	English Witchcraft Trials	176
	The Victims of the Great Witch-Craze	181
	An Organization of Witches?	186
	The End of the Witch-Craze	189
	Continuing Popular Beliefs in Witchcraft	191
11	**Our Contemporary Witches**	199
	Witch-Beliefs and Social Change	200
	The 'Witches' of Contemporary Industrial Society	205
Glossary		213
Bibliography		216
Index		226

Preface

This book is an introduction to the study of the witch. Although primarily an anthropological work that incorporates history, it is written in a style that is intended to make it of interest to as wide an audience as possible. This is because I believe its topic has social relevance to the present time. I try to demonstrate this by relating the witch beliefs of less complex societies to the political use of scapegoats in our own time, with its ominous implications for our immediate future. I have written for the layperson and the undergraduate student,[1] but I hope the book will be useful for professional anthropologists and historians. I believe they will find its wide selection of case material useful for their teaching, and possibly for their research. They may also find some of my hypotheses stimulating. I have tried to write a book that can be enjoyed by the general reader, but I am hopeful that the historian and anthropologist will find it a worthwhile contribution to their literature.

Earlier drafts have been read by lay persons, students and academics. I wish to thank the following for their helpful criticisms and comments: Mr Harry Barton, writer and playwright; Mr Dominic Bryan, D.Phil. student Anthropology/Sociology and Research Assistant for The Cultural Traditions Group, Community Relations Council, Northern Ireland; Dr Simon Harrison, Reader in Social Anthropology, University of Ulster at Coleraine; Dr Keith Lindley, Senior Lecturer in History, University of Ulster at Coleraine; Professor Peter Morton-Williams, Professor Emeritus, Sociology, University of Ulster; Mr Alexander Pope, retired shoemaker; Mr David Pownall, playwright and writer; Ms Mary Reid, D.Phil. student, Anthropology/Sociology, until recently a Community Development Officer with the EEC, Ireland (Donegal); Mr Bryan Rooney, Head of the

---

1. Because I have not written only for the expert, some classic anthropological works are omitted from my examination. For example, I have not discussed Mitchell's *The Yao Village* (Mitchell 1956) or Middleton's *Lugbara religion* (Middleton 1960). For my purposes, Marwick's data on the Cewa have proved sufficient. It is not that this is superior to the other two monographs; simply that the aims of this book made only one of these works necessary, and circumstances made it the Cewa one. Again, because I am not producing a work solely for professional anthropologists, and because my focus is the witch and not witchcraft as such, I do not go into any detail on topics such as lineage fission (Mair 1969: 116–38) or some of the less general, structural aspects of witchcraft accusations. They are not germane to the points I wish to make.

Preface

Department of English, Dominican College, Portstewart; Dr David Sturdy, Senior Lecturer in History, University of Ulster at Coleraine.

Professor Mary Douglas read an earlier draft Introduction and commented on the Bibliography. Mr N. N. Dodge, Lecturer in Sociology, University of Ulster at Coleraine, commented on the sociological theory. I particularly wish to thank Dr Rosemary Harris, Reader in Social Anthropology, University College London, for her meticulous reading of an earlier draft of what later became Chapters 2–8. None of these persons bear any responsibility for any of the statements herein. My thanks also to Edwin, David and Patrick Sanders, who did much of the word-processing.

Because I hope to reach a lay audience, I have reduced the use of anthropological terminology as much as possible. Where it cannot be avoided, I have provided definitions of terms which I have tried to make as simple as possible without losing their essential meaning. They are contained in a glossary that is placed after Chapter 11. (There are short sections of a more theoretical nature at the ends of Chapters 3 and 5. The general reader may wish to pass over these to the following chapters.) In seeking to write for the non-specialist, I have found Professor Lucy Mair's book on *Witchcraft* (Mair 1969) a useful guide. As far as I am aware, Professor Mair's is the last anthropological book on witchcraft aimed at a general readership. As well as bringing this situation up to date, my book differs from hers in its focus. It deals with the witch, as a person and as an idea, and not with the general topic of witchcraft. It also draws its examples more widely. Mair was concerned mainly with African data. Although she incorporates some European material, it is contained in a single chapter that is almost an addition to her book. I have taken my anthropological examples from a much broader field, and have sought to integrate Europe more fully into the main themes of this work.

There are other books which bring together anthropological and historical material. Keith Thomas' *Witchcraft and the Decline of Magic* (Thomas 1973), first published in 1971, applied hypotheses from the anthropological study of witchcraft to English data from the sixteenth and seventeenth centuries. Alan Macfarlane's *Witchcraft in Tudor and Stuart England* (Macfarlane 1970a), did so to an even greater degree. However, both these works are concerned primarily with the history and sociology of English witchcraft, and both were written at the start of the 1970s. Again, they deal with the more general topic of witchcraft, whereas the present study is concerned primarily with the witch.

Writing about witches poses problems of presentation. In many of the societies I discuss, life has changed since the cases were collected. Some of the examples are taken from works published several decades ago, and witchcraft beliefs may have altered or become less important in the

Preface

societies concerned. Often I do not know what changes have taken place. Under these circumstances, to write in the present tense, as though the situation recorded still pertains, may not only be misleading; it may give offence. Consequently, I have chosen the admittedly arbitrary procedure of using the past tense for data collected prior to the 1960s, unless I have information indicating that the situation regarding witchcraft remains relatively unchanged. Correspondingly, data collected between the 1960s and the present are discussed in the present tense.

Gender presents another problem. Most Western European witches were women. Consequently persons of Western European background tend to think of a witch as 'she'. But in some societies witches predominantly were men. In others they might be drawn in significant numbers from both sexes. English has no special term which means a member of either sex; it uses 'he'. I have sought to avoid confusion as much as possible by making it quite clear when I am talking of societies where witches are of one sex only, and when they may be of either. But it has not always been easy to avoid clumsy phraseology in order to do so. Sometimes I have resorted to 'he' when I mean a member of either sex.

I wish to acknowledge permissions for quotations from the following works. A quotation from E. E. Evans-Pritchard, *Witchcraft, Oracles and Magic among the Azande* (abridged and with an Introduction by Eva Gillies), 1976, by permission of Oxford University Press. Quotations from 'The diary of Antera Duke', in Daryll Forde (ed.), *Efik Traders of Old Calabar*, Oxford University Press, 1956, by permission of the International African Institute. Four quotations from Clyde Kluckhohn, *Navaho Witchcraft*, 1967, by permission of Beacon Press. A quotation from Keith Basso, *Western Apache Witchcraft*, 1969, by permission of the University of Arizona Press.

Finally, I wish to thank the University of Ulster at Coleraine and the Department of Sociology, University of Ulster, for a term's leave of lectures during 1993, which helped me to complete this work, and Mr Arthur McCullough, head of the Department of Sociology, for his encouragement and support throughout.

–1–

Power and the Witch

In Europe today, once again the witch-finders are on the march. (In much of the Developing World, with its continuous social and economic crises, they have been active for a generation or more.) World recession and rapidly changing technology are creating rising unemployment and threatening established occupations and social statuses. At the same time there are record rises in crime statistics and terrifying new threats like AIDS. Added to this, the collapse of the Soviet Union has created new social and political problems in much of Europe. All these factors are working together to create at the end of the twentieth century, with its rapidly changing present and its uncertain future, a climate particularly suitable for the development of hunts for scapegoats to blame for these conditions and anxieties – hunts for the witches of the turn of the twentieth century and the beginning of the twenty-first.

Europe's witch-hunters have been relatively quiescent since the Second World War, but once more there are powerful moves to seek out the persons supposed to have caused her problems. They are focusing their attention upon minority groups, and in particular upon racial and ethnic minorities. These are in danger of becoming Europe's witches of the late twentieth century. It is claimed that their numbers, behaviour, and attitudes have created social problems and are threatening the whole social order. Many minority communities are stigmatized as 'immigrants', even though they may have resided in their host countries for two generations or more. In some European countries this 'witch-hunting' is associated with a general xenophobia against foreigners, immigrants and asylum-seekers.

This book examines the idea of the witch, the person whose character is believed to be such that he or she causes personal and social misfortune, often deliberately, and whose life is said to be a perversion of everything that is proper or moral. It does so by examining the findings of anthropological and historical studies of witch beliefs and accusations. However, I would like to think that it has a message, and a warning. If we can understand the witch-phenomenon – that is, if we can understand what kinds of people are accused of monstrous, mythical offences, and why – then we may be able to counter the witch-belief. In the process, we may safeguard the rights and welfare of minorities. At the least, we should have understood some of the reasons why minorities are persecuted. If we do not do so, then we might expect their persecution to increase, with all the destructive potential that this holds for society. By examining the witch from pan-cultural and historical perspectives, not only may we learn something about our history, we may become aware of disturbing aspects of our contemporary society. Whilst I do not believe that awareness alone is sufficient to prevent such persecutions, if it becomes linked to political activities and groups then it may help to do so.

My examination is done through the use of case studies. I examine cases of individuals who are accused of witchcraft. They are taken from a variety of societies, from several continents, and in the case of Europe, from a long span of history. I hope that the arguments of the book speak through the examples, and that in this way the reality of what we are discussing becomes more obvious and dramatic. Occasionally a case is quoted verbatim from the narrator, who is one of the participants, and this adds to its drama.

The Cases

In witchcraft cases, as in other social situations, persons concerned with the activities, be they participating, observing, or simply hearing about them or relaying accounts of them later, have their own interpretations of events and apply their own meanings to them. When they are discussed they are recalled selectively and reinterpreted. Perceptions, interpretations and recollections are not simply idiosyncratic, they are influenced by social factors such as group membership and the nature of interpersonal relationships. As we shall see, the resulting differences in interpretations are a crucial aspect of the operation of witch-beliefs.

In general, the non-European cases of witchcraft used here are based upon data compiled by anthropologists. As with all field material, these data are influenced by the interests and theoretical concerns of the compiler, so that they must be regarded as his or her particular interpretation of a happening. Sometimes the anthropologist was present when the events described were taking place. Where this was so, or when they took place in the relatively recent past, a variety of techniques will have been used to ascertain the sequence of events and how they were interpreted by different parties at the time. Consequently, we may expect the case to contain a relatively reliable account of events as they occurred, and of how they were interpreted initially. It may be possible to demonstrate the process whereby events were reinterpreted over time – an extremely important process in the operation of witchcraft beliefs. The further anthropologists are removed from the data they are studying, in location and in time, the more difficult are the problems of ascertaining events, and the more cases become the interpretations of the different parties involved, or of their descendants.

All case studies pose these two sets of problems of interpretation: the problems of ascertaining events – what happened and how people behaved and what their interpretations were at the time, and the problems of the meanings of the events to different participants – who made a particular interpretation of events, and when and why? They usually become more acute as we move from data collected from a contemporary situation, or about the recent past, and into history. Historians cannot immerse themselves in the cultures of past periods the way the anthropologist may do in a study of the present. They cannot have direct access to the pattern of meanings that constituted the culture of a past society, and often they do not know how different individuals interpreted a particular incident. Frequently they cannot know how typical any recorded occurrence, such as a local witchcraft accusation, is of the time. Many of their data are from written records that are partial and represent the interpretations of particular individuals, professions, and classes. Their facts have already been selected and interpreted by particular social groups of the period concerned, and further selected by the vagaries of the survival and availability of records. To this material the historian has to bring his own interpretation, which is influenced by his philosophy and by the perceptions and events of his time (Carr 1987: 7–30).[1] These problems may be particularly acute with respect to folk or popular culture, which usually received less attention from contemporary recorders and was interpreted through the eyes and values of the educated. Folk culture is important for an examination of witch beliefs, and is a field to which many historians are now turning.

The Navaho witchcraft cases that are used here illustrate some of these problems. Most of them constitute an individual's recollections and interpretations of a past event, or are hearsay stories. Consequently they have to be treated as illustrations and examinations of Navaho beliefs about witches and methods of witchcraft detection, and not as accounts of specific events.

Given the problems of interpretation, we must approach our cases with caution, particularly with regard to their reliability as records of actual events. However, cases provide the examples that are the basis of any sociological form of study, and increasingly they are compiled by researchers who are aware of these problems. We can use them to make

1. For a debate on the aims and methods of history and anthropology, see Geertz 1975; Thomas 1975.

at least some generalizations about the witch and culture and society. At the very least they demonstrate the cultural and social themes that various peoples have expressed through the idea of the witch.

When we examine witchcraft in the local community in Europe, we may hazard the assumption that there are likely to be similarities with witchcraft beliefs and behaviours revealed by anthropological studies of local communities in other types of society, because the communities under investigation shared certain basic features. In fact, anthropologists and historians have reached many similar conclusions when discussing witchcraft in the local community, in Europe or elsewhere. Sometimes this is because historians have used anthropological hypotheses.[2] Others, who have not, have still produced conclusions on, for example, patterns of witchcraft accusations, that support anthropological findings.[3] In addition to the case studies, I have included much information from the findings of both anthropologists and historians concerning witchcraft. As well as suggesting many remarkable similarities in witchcraft beliefs and activities between different cultures anthropology and history also show that there are significant differences, which often result from cultural and social factors that can be revealed by the kinds of cases available to the European historian. These are cases that involved and interested those European groups that compiled records.

In some of the anthropological cases the data were collected several generations after the events took place. They constitute ethnohistory. The ethnohistorian may have greater access to the cultural meanings of his data than does the historian of Late Medieval and Early Modern Europe.[4] The lives of his informants may involve cultural meanings similar to those of the period under study. In some cases the historical events were also recorded by literate contemporary European chroniclers. One case (case VIII-5) is a diary kept by a native recorder in the eighteenth century. Here the anthropologist, like the historian, has to interpret the meaning of records written many generations before his time.

Conflict Theory

Some influential historians argue that anthropologists have little or nothing to offer them in their examination of witchcraft in post-medieval Europe. Professor Trevor-Roper, who initiated the recent historical study

---

2. For example, Clive Thomas (Thomas 1973) and Alan Macfarlane (Macfarlane 1970a).
3. For example, E. William Monter (Monter 1976) and H. C. Eric Midelfort (Midelfort 1972).
4. 'Early Modern' is the historian's term for the period in Europe from about 1450 to about 1720 (Clark 1966).

of witchcraft, in his *The European Witch-Craze of the Sixteenth and Seventeenth Centuries* rejects any possible anthropological contribution towards his study (Trevor-Roper 1965: 9; 1988: 9). Europe's history is unique, and cannot be illuminated by the cross-cultural study of local communities and different types of society. More recently, E. William Monter and Christina Larner both have denied a role for anthropology in the study of Europe's demonological witch-belief and her sixteenth- and seventeenth-century witch-hunts (Larner 1981: 26; Monter 1976: 101). Monter claims that cross-cultural comparison of the kind used here is 'useless for understanding continental European witchcraft'. Witchcraft in non-European societies and classic European witchcraft are totally different phenomena (Monter 1976: 10–11). Larner says that her interpretation of the European Witch-Hunt and its Scottish version 'rests essentially on themes of political sociology: power; dominance; ideology; and legitimation', in contrast to anthropological theory (Larner 1981: 192).

I would argue that it is precisely *because* anthropology also is concerned with the themes of power, dominance, ideology and legitimation that it can make an important contribution to the study of the witch, including the demonic witch of post-medieval Europe. In my study I have adopted a 'Conflict Theory' perspective (Alexander 1987: 127–55; Collins 1994: 47–111; Craib 1984: 59–70). I believe that competition for power* provides the dynamic of society. It is an important determinant of the form of society and a major factor promoting social changes. My approach derives particularly from the ideas of the classical sociologist Max Weber. Society consists of individuals and groups that are acting in pursuit of what they perceive as their interests, and this brings them inevitably into competition with other individuals and groups. Interests vary with culture and with the form of a society. Cultures set goals, and evaluate ways of attaining them, but interests are not simply a reflection of cultural standards and values. What individuals see as their interests are the result of their perceptions of themselves, of others, and of their society, and these are influenced by social factors such as social position and group membership (Marshall 1994: 253–4).

Power can be derived from a number of sources. Examples include the occupation of a culturally recognized ascribed status carrying rights over persons designated as dependants; control of esoteric knowledge, such as certain types of religious knowledge; control of the use of economic resources; and control of physical force. Access to one source often leads to control of other sources, and power is often derived from

---

* Sociological definitions of terms such as 'power' and 'interests' are contained in the Glossary, situated after Chapter 1.

a number of sources simultaneously. In the most complex societies, control of economic resources and of physical force tend to be more prominent power bases than they are in less complex societies.

People seek power in order to achieve and advance their interests, and organizations seek power in order more effectively to pursue their members' interests, or the interests of influential groups and individuals within them. Just as there is competition between individuals and groups, so is there competition within groups as factions and individual members seek to advance their own interests, although organizations and groups may demonstrate consensus and solidarity in their competition with others in order to be more effective in pursuit of their aims. Powerful individuals within organizations identify the interests of the organization with their own, and organization interests and personal interests become modified in the process. Society may be viewed as constituting a series of levels, within each of which there is competition for domination by individuals, factions and groups. In less complex societies the competing units are individuals, families and kinship groups. In more complex societies, as well as individuals, there are many competing organizations and groups, such as social classes and élites, interest groups, and institutions such as churches, political parties and economic corporations, each with its own internal factions.

Pursuit and exercise of power generate conflict, as those with less resent its exercise by those with more, and seek to improve their relative power positions. Those with power seek to legitimize its possession and transform it into authority – socially accepted power – in order to ensure their position. They do so by making some kind of ideological appeal, such as an appeal to established cultural standards and values (Marshall 1994: 411–13). If conflict is not actively being expressed, often this is not because people and groups are satisfied with their power position. It is because they see themselves as unable, or are unwilling, to contest it.

My argument concerning the witch is as follows: As all anthropologists recognize, certain categories of persons are particularly open to accusations of witchcraft because of their social position and their relationships with other persons. However, the members of only some of these categories are likely to have the witch-label stick to them. Their relationships with the community are of a kind that is likely to arouse suspicions of witchcraft among large numbers of its members. (I argue that these are not the categories usually stressed in social anthropology, which has tended to over-emphasize accusations resulting from contradictions in certain kinds of interpersonal relations.) These persons are the witches of tribes and other small-scale societies and communities. In these societies accusations of witchcraft are often treated as a personal matter involving individuals or small groups, but when someone becomes widely stigmatized as a witch, his or her actions are condemned by the whole community.

When witchcraft occurs in situations involving control of, or competition for, a marked degree of power, the nature of witchcraft, and hence of the witch, undergoes a change. Where accusations are associated with positions of real power and influence, witchcraft is treated as a very serious offence. It is held to cause widespread misfortunes, and the emphasis is on punishment of the guilty, rather than on reconciliation between the parties directly involved or on severing the relationship between them. Instead of being allowed to live, witches are likely to be killed. The process of a person's acquiring the status of witch, that normally requires a long period of labelling often taking many years, may be dramatically shortened, as it is now imposed from above, by political power or authority, and is not a gradual development from within the community. Persons so designated have the witch-image, the set of cultural beliefs about the antisocial and abnormal behaviour of witches, applied to the interpretation of their everyday behaviour, whereas in other circumstances persons accused, even successfully accused, do not have their behaviour interpreted in this way by many members of the community.

Witchcraft may be directly involved in situations of power, when important individuals are its victims or are associates of its victims. This is the case when witchcraft attacks a king. Or the involvement may be indirect, a powerful party seeking to circumscribe witchcraft as part of its competition for power, but not itself being a target of witchcraft. In Western Europe witchcraft became defined as a major public offence as a result of the medieval Church's conflicts with religious groups that it branded as heretical. In Early Modern Europe political regimes persecuted witches in order to demonstrate their legitimacy and their control of the legal process. When small-scale societies are undergoing drastic, unpopular change, religious movements may arise whose leaders may blame the new social problems on witches. In contemporary industrial societies particular groups may be scapegoated to explain social problems, but since these societies have more secular ideologies, the 'witch', while endowed with mystical attributes, is not portrayed as overtly magical. All these situations involve the use of witchcraft accusations by powerful individuals or groups.

In my examination, I have found anthropology particularly valuable for illuminating why certain categories of persons attract accusations of witchcraft. History shows us the important and varied role that political factors can play in witchcraft accusations. Far from being opposed, I would argue that the studies of social and cultural anthropology and those of history complement one another.

A Deed without a Name

Anthropologists have studied witch-beliefs in a wide variety of cultures and societies, and this helps us to define the kinds of persons who are accused of witchcraft. Historians have made detailed studies of the development of the witch-belief in medieval and post-medieval Europe, where the idea of the witch is most likely to be familiar to my readers as that of their childhood fairy tales and of films like 'The Wizard of Oz'. The anthropologist's studies place European witches in a wider context. They show why witches were drawn from certain types of persons. The historian adds to the anthropologist's findings, and illuminates the way witchcraft beliefs develop when they become associated with politics – with the pursuit of power. Both reveal that the labelling of people as witches is not random. Only certain categories of people are in much danger of being so stigmatized. Knowledge of how witches were selected in non-industrial societies helps us to have an understanding of the selection of the more 'secular' scapegoats of our own time: Hitler's selection of the 'Jews' and other minorities; Stalin's search for 'Trotskyites'; McCarthy's search for 'communists'. It may help us understand much of the current European political concern with 'immigrants'.[5]

Industrial societies have their own witches, and may develop their own witch-hunts. When I say this, I am not talking of those few persons who believe themselves engaged in certain kinds of supernatural practices, either as Satanists or as followers of what they believe to be some pre-Christian fertility cult. These do not belong to the study of the witch as understood by historians and anthropologists.[6] They are not analogous to witches in other societies, as they are not accused of causing *misfortunes*. If they are accused of anything, it is usually of carrying out certain offences, often of a vile nature such as child abuse (see La Fontaine 1994) or ritual murder.[7] Consequently they are not discussed here.[8]

I am talking of those groups whose punishment may be sought in a misguided attempt to cure our contemporary ills – the members of minority groups, or persons believed to hold certain opinions, who may be punished cruelly in ways reminiscent of the treatment of the traditional

---

5. I place these terms in quotes, because in these examples many of those persecuted could not reasonably be classed as members of the population named.
6. Briggs (1989. 19–20) refers to them as 'pseudo-witches'.
7. It is currently (October 1994) being suggested that this may have been an aspect in some of the 1994 deaths of the members of the Order of the Solar Temple in Switzerland.
8. Good anthropological or sociological analyses of contemporary groups who call themselves 'witches' have been made in England by T. M. Luhrmann (Luhrmann 1989) and the USA by Marcello Truzzi (Truzzi 1972), and there are recent works written from a popular historical or a social standpoint such as those by Jeffrey Russell (Russell 1981) and Peter Hough (Hough 1991).

witch, and for as mystical a crime. Our modern witch-hunts often have been far more horrendous than those of the much-discussed, and apparently much-exaggerated,[9] 'European Witch-Craze'. The social tendency to seek out 'witches' as a cause of misfortune remains as active in Western Europe at the end of the twentieth century as it was in the sixteenth; and as it was in tribal societies and as it is in some developing countries.

---

9. See Larner 1985: 26, 30, 36–7, 71–2.

–2–

The Witch and Society

In his examination of the development of the witch-belief in Western Europe (Cohn 1976) Professor Norman Cohn cites an early sixteenth-century case of a woman accused of witchcraft in the Swiss canton of Lucerne. A summary is given below:

Case II-1

Stürmlin was a 'wise woman', a practitioner of benevolent magic, in the village of Rüti in the Willisau. She had picked out a girl who she hoped would marry her son, but she married another man, Sebastian. Sebastian proved impotent with his new wife, and blamed Stürmlin. He accused her of witchcraft in 1531. In his deposition he said that she would often enter his house unannounced and leave without saying a word. When he and his wife took a bath together in the hope that this would cure his impotence, she appeared and warned that the bath might prove too strong for them. This so terrified Sebastian's wife that she experienced violent cramps. When he forbade Stürmlin entry to his house his cattle died and his horses over-ate and were unable to work. He claimed that in church one day Stürmlin gave him a terrifying look, and later that same day he developed an intense pain in his neck.

Sebastian reproached Stürmlin for bewitching him, but she denied it and asked him not to slander her. She said she was trying to help both him and his wife with her prayers. Clearly Sebastian and his wife had worked themselves into such a hysteria over Stürmlin that any contact with her caused them to suffer psychosomatic disorders (Cohn 1976: 243).

In this brief example we have many of the features that characterize the witch world-wide. Stürmlin is believed to be a malicious person who uses her powers out of spite to harm those who offend her. (Consequently, she is accused of witchcraft by someone who believes he has given her offence.) She behaves in what is considered to be an odd manner – she enters Sebastian's house and leaves without saying a word. Once she is suspected of witchcraft, her 'victims' increasingly blame her for their personal misfortunes, until a whole series of ills have been laid at her door – Sebastian's impotence, the deaths of his cows and the incapacity of his horses, the pain in his neck. In fact she becomes such an obsession that fear of her alone is enough to produce illness. Once a misfortune is attributed to her malice, comments she has made, possibly with the most innocent of intentions, become reinterpreted as threats – for example, that the bath might prove too strong for Sebastian and his wife. Stürmlin is also a type of person that is often at risk of suspicion of witchcraft in this particular community. She is a woman, and in European witchcraft the suspect commonly is a woman. Almost certainly she is elderly, although we are not given this information. She has the power to cure illnesses and other misfortunes, and the same power that she uses to cure them might also be used to cause them.

Witches then, are persons with an ability to harm by mystical means. Within a culture there may be several types of persons with this ability, from priests to shamans and sorcerers. What distinguishes the witch is that he or she is motivated by hatred or malice, or simply by enjoyment of causing harm.[1] Primarily, witches do not use their powers for social ends or to improve their personal circumstances. Whereas other unsocial users of mystical power may be motivated by selfishness, or greed, or ambition, or by similar personal motives, witches act because of their malicious natures. Witches are evil. Consequently they carry out their nefarious activities hidden away from human sight. Often they perform them when they are hidden by the dark. 'Normal', ordinary people find their behaviour incomprehensible as well as terrifying. Their activities oppose all that is valued in a culture, and when they meet together they create a society that is a grotesque caricature of the human society within which they operate, that they attack from within. Shakespeare aptly summed up their activities when, in Macbeth, he described them as constituting 'a deed without a name' (*Macbeth*, Act 4, Scene 1).

The belief in witches has been found in most societies, where they are used as an explanation of personal and group misfortunes. They are never the only form of mystical agent used to explain such misfortunes. Among others may be cursing, the anger of ancestors, or the malevolence of ghosts (Mair 1969: 11–14),[2] and people often examine a variety of possible explanations before deciding upon witchcraft (Macfarlane

---

1. This was also the opinion of European believers in witchcraft – whether they were educated judges and demonologists, or illiterate accusers, or 'witches' making confessions (Briggs 1989: 34).
2. For general accounts of different types of mystical agency, and discussion of whether it is possible to relate different kinds of mystical beliefs to different patterns of social relationships, see Gluckman 1965: 223–9; Lewis 1976: 93–105.

1970a: 108). Different explanations are likely to be invoked in different sets of circumstances, and different parties may attribute different causes to the same event. In sixteenth- and seventeenth-century England, for example, close relatives of a deceased person were more likely to invoke witchcraft as cause of death than were other members of the community, because they were more involved in his or her personal relationships and antagonisms. Other persons often had a less biased attitude. Consequently, it was more common for relatives to make their suspicions public when the person they suspected already had a local reputation as a witch. Otherwise, knowing that they were unlikely to be believed, they might keep their suspicions to themselves (Macfarlane 1970a: 179).

A great number and variety of cultures have a belief in witches, and to demonstrate this I give four cases that involve witchcraft beliefs, from four widely different types of society in different parts of the world. Our first three examples are from societies in which belief in witches is an established part of the culture. They are the Paiute of the arid Great Basin region of western North America; the Kaluli of the tropical forest of New Guinea; and the Shona of contemporary Zimbabwe. The fourth is from a society where the great majority of persons no longer believe in witches – England in the middle of the twentieth century.

Our first example is from the Northern Paiute, who until the second half of the nineteenth century were semi-nomadic food collectors, hunters, and fishermen (Fowler and Liljeblad 1986). They occupied a vast and often inhospitable area of Nevada and Oregon, living in small camps that usually contained two or three families. Twice a year these camps united with others to form larger groups, and a typical winter camp consisted of about 50 persons.

Among the Northern Paiute any person could seek mystical power. Usually it came unsought in a dream, but it might be inherited from a close relative, usually deceased, or deliberately sought in certain caves in Northern Paiute territory. Those few persons, of either sex, who acquired sufficient power to enable them to do good or harm to others became shamans. Persons who had bad mystical power could use it to make others ill or to kill them, and consequently they were called 'people eaters'. Bad power often passed down through families. The possessor could cause harm by intentionally thinking bad thoughts or by 'throwing' the power into others, and could cause illness and death unintentionally through malevolent thoughts. Sometimes the victim dreamed of the witch, providing a clue to his or her identity. The cause of an illness was diagnosed by a shaman, who attempted to cure the sufferer (Fowler and Liljeblad 1986: 450; Whiting 1950: 28–33).

Beginning in the 1840s, Paiute–Euro-American contact precipitated a number of conflicts, most of which were over by 1869. The first Paiute reservations were established in 1871, with the intention that they would settle there and become farmers, and in the late 1870s and early 1880s day schools were established on Pyramid Lake and Walker River reservations (Fowler and Liljeblad 1986: 455–61). However, the settlement policy initially met with only limited success because the areas chosen were unsuitable for farming. Our example comes from this period.

Case II-2

In 1882 Old Winnemucca, the well-known leader of one of the Northern Paiute bands, became very ill. He was over 90 years old, and had recently married a young widow with a child. A Paiute shaman worked all night to cure him, but he had lapsed into a coma from which he would occasionally awake. Once when he did so he said he had dreamed that his wife had bewitched him. The Paiutes believed her guilty of her husband's illness, and she became increasingly frightened about what they would do. She was taken to a spring to remove the effects of her witchcraft by bathing (see Whiting 1950: 34, 51), but when she was unguarded she found a rope and attempted to hang herself. This confirmed her guilt. The Paiutes held a council and decided she must suffer the customary fate of a witch – to be stoned to death and her body to be burned. Her child must die with her, as it would inherit her witchcraft. She was taken back to the spring and forced to bathe, and then sprinkled with ashes. When night fell she was tied by the leg to a tree stump, enclosed by a circle of fires. The Paiutes circled her, chanting. One stepped into the circle and started to harangue. When he had finished he seized the child from its mother and smashed its head against a rock. The Paiutes recommenced circling and chanting, until the leader picked up a rock and struck the woman. She was stoned to death, and the corpses of mother and child, along with their belongings, were consumed on a fire.

Winnemucca was now expected to recover. Instead he lingered until 21 October, when he died (Canfield 1983: 196–9).[3]

Our second example is from the Kaluli of the Great Papuan Plateau of New Guinea. The Kaluli are subsistence agriculturalists who live in tropical forest. The typical Kaluli village consists of a single longhouse containing about 15 families (about 60 persons). The men are usually from two or three different patriclans, which have members in other longhouses. The Kaluli population has been falling since the late 1940s

---

3. Much of Canfield's account is taken from contemporary newspapers of the period. For a first-hand account of another witch-killing from the Northern Paiute for the same period, see Scott 1966: 65–74, the case of Annie Doughnuts.

A Deed without a Name

owing to high infant mortality and a series of devastating influenza epidemics. Australian administration began to be effective from the end of the 1950s, and revenge raiding was ended in 1960 (Schieffelin 1977).

Among the Kaluli all deaths are due to witches, who are men (or, rarely, women) who, unknown to themselves, have an evil aspect, a 'witch child' (Schieffelin 1977: 101), in their heart. When its possessor is asleep it takes his form and creeps around at night looking for victims. It can only be seen by a person who is seriously ill or by a medium (Schieffelin 1977: 78, 98, 101, 127).

Case II-3

In 1954 sickness was prevalent among the Kaluli. One of those who died was Gisa, the wife of Wanalugo. Before she died she said she was bewitched by Hagabulu, a close classificatory mother's brother of her husband. Another woman who died had told her husband that she was bewitched by a man called Sidawɔ. Sidawɔ was ignorant of the accusation, and when he came to the funeral of Wanalugo's wife he was killed as he walked through the door of the longhouse. This made Wanalugo determined to kill his wife's killer, despite the fact that the kinship relationship between them was so close that the homicide would be considered a particularly horrible one. He was encouraged in this by his friends.

Wanalugo scouted the vicinity of Hagabulu's longhouse. While he was doing so he met two men of the longhouse, Ola and Hæmindæ. They were not members of Hagabulu's clan, so he told them his intention. They agreed not to warn Hagabulu. Ola was not unhappy to see Hagabulu killed, because when his classificatory brother was accused of witchcraft Hagabulu had killed him.

Wanalugo asked two men, Bugamia and Meyɔ, to visit Hagabulu's longhouse in order to admit the attacking party. Then he and his friends gathered their supporters to form a raiding party of about thirty-five persons. They met in the dark in the forest, and Wanalugo told them who they were going to kill. Until then he had kept the destination secret from all but a few, in order to prevent Hagabulu being warned. Persons with close ties with the alleged witch or with relatives in his longhouse had not been informed and were not included in the raiding party.

At dawn the party surrounded Hagabulu's longhouse, and at a signal from Wanalugo Meyɔ opened the door to admit them. Bugamia seized the wakening Hagabulu and Wanalugo struck him with his club, but Hagabulu was an experienced fighter, and he ducked and received only a glancing blow. Hagabulu dived over the partition into the women's section of the longhouse dragging Bugamia with him, and Wanalugo followed and killed him. There was much confusion and fighting in the house, but of the nine men of the longhouse, three had been secretly warned, and although they made much noise they did not fight.

Hagabulu's body was thrown off the veranda and the raiding party retreated to the longhouse gardens. Someone shouted that they should all stop fighting as it was getting light. A member of the raiding party cut open Hagabulu's body and took out his heart, examined it and proclaimed it the heart of a witch. Fighting was suspended, and Tiago, a man of the longhouse who had been warned of the raid by Ola, came out and confirmed that it was indeed the heart of a witch (which is soft and yellow.) At this the longhouse men declared that they would not fight further, and the raiding party left for Wanalugo's longhouse taking Hagabulu's body. On the way the body was cut up and the meat divided to be eaten. (Members of Wanalugo's longhouse would not eat any of the meat for fear that another witch, angry at the killing of his brother, would revenge himself on a member of their community.)

At Wanalugo's longhouse Hagabulu's heart was displayed on a stake, and the people barricaded themselves in to await the arrival of his relatives and supporters to demand compensation for his killing. However, sickness continued, and a youth of the longhouse died. This was interpreted as retribution by some other witch for Hagabulu's death. The men of the longhouse were angry, and consequently Hagabulu's supporters decided to stay away.

For more than a year there was no communication between the two communities. Then a woman of Wanalugo's clan who was married to one of Hagabulu's brothers came to Wanalugo's longhouse with her two children. She said she was tired of not seeing her relatives. The people of the longhouse reciprocated, giving her some items of wealth to distribute to those persons most upset over Hagabulu's death. The longhouses resumed friendly relationships, and soon a clan from Hagabulu's longhouse gave a woman in marriage to Wanalugo's brother, who had not participated in the raid, and this was regarded as cementing the peace (Schieffelin 1977: 80–7).

Our third example is from the Shona of Zimbabwe. The Shona are by far the largest ethnic group in Zimbabwe. In rural areas they live in small family hamlets or villages comprising a number of families usually linked by agnatic ties between men. A family occupies a number of huts, and has a common living hut in which pre-adolescent children sleep and which contains the shrine to the family ancestors. Most of the men of a village are of the same patrilineal clan, and their wives are members of other clans. A village has a headman who is responsible to the clan chief or to a clan subchief. A subsistence economy is practised, based on millet or maize, but as with other southern African Bantu peoples migrant wage labour is an important factor in the rural economy and a strong influence on rural life, with men leaving tribal areas to move to mines, towns and European farms for periods of wage-work (Crawford 1967: 9–39; Gelfand 1967: 6–10). This was the situation as described in 1967.

The Shona witch generally inherits his or her power from a spirit, usually from a recently deceased ancestor who was a witch (generally a parent or grandparent). The spirit appears to the potential witch in a dream and gives instruction. Sometimes, however, an individual obtains witchcraft from a living witch. Witches are usually women and usually adult, although occasionally a child may be a witch. The witch may visit her victim at night in spirit form, leaving her material body sleeping in her hut. She can cause harm in a number of ways: shooting a magic projectile into her victim; using poisons, medicines, or charms;[4] or sending an animal familiar to bewitch on her behalf. Witches and witchcraft are detected by a traditional doctor (*nganga*). People are protected from witchcraft by their ancestral spirits, and they can be harmed by witches only if they offend these spirits so that they withdraw their protection (Crawford 1967: 107–26, 161–221; Gelfand 1967: 11–61).

Case II-4

In December 1958 (when Zimbabwe was still the British colony of Southern Rhodesia) it was reported from Bikita, about 200 miles from Salisbury, that three women, Mazwita, Puna, and Netsayi, had confessed to their community that they killed Mazwita's husband, Mukhozu, by witchcraft. From the accounts of this case (Crawford 1967: 49–56; Gelfand 1967: 47–8), they appear to have taken pride in describing their prowess as witches. They said that in September 1958, at one of their nightly witchcraft sessions, they decided to poison Mukhozu because he favoured his other wife and would not buy clothes for Mazwita. Each woman provided some of her 'medicine', which they mixed together in a calabash with a little beer and poured into the sleeping man's mouth. Mukhozu became ill. He moved to a different hut, an indication that he believed himself bewitched. He moved again, but did not return to his wife's hut, which indicated that he believed his illness originated there. When he became very ill Mazwita asked the village headman to call a meeting of the villagers. She told them that she and Puna and Netsayi had bewitched Mukhozu, and they would make a thin porridge with which

---

4. Individuals who obtain poisons or medicines from an evil doctor in order to harm another are also called witches (*muroyi*), as is such a doctor, but are less feared than the hereditary or 'night witch' (Gelfand 1967: 11–41).

to cure him. (Mazwita told the police that she confessed because she regretted what she had done.) Puna and Netsayi at first denied being witches. Then they admitted it, and each of the three described the medicine she had supplied to bewitch Mukhozu. In spite of his being given the porridge, Mukhozu's condition became worse and he was taken to a clinic, where he died.

In their statements to the police the three women boasted of their witch-activities (Crawford 1967: 51). Netsayi said she had been friendly with Mazwita and Puna for two years. Early in 1958 Puna asked her, in Mazwita's presence, if she would like to be able to bewitch people. She said that she would, and all three met naked that night at a place designated by Puna. Puna rubbed an ointment on Netsayi's face to enable her to accompany them on their night-time excursions.

The three women were committed for trial at Bikita on 10 February 1959, and subsequently indicted before the High Court on a charge of attempted murder, or alternatively of contravening Section 7 of the Witchcraft Suppression Act. They were acquitted 17 March 1959.[5]

Our fourth example is from a society where beliefs in witches have become socially unimportant, and are held only by a minority. It is included to show that such beliefs are not the prerogative of technologically 'unsophisticated' societies.

Case II-5

On 14 February 1945 Charles Walton, an elderly labourer from the English village of Lower Quinton, near Stratford-upon-Avon, was discovered murdered on nearby Meon Hill. He was pinned to the ground by a pitchfork, and a cross had been carved on his neck and torso with a billhook. A few days later the body of a black dog was found hanged on Meon Hill.

The murder was remarkably similar to one in 1875 in the neighbouring village of Long Compton, when an old woman, Ann Tenant, had been killed in similar fashion. Her murderer claimed she was a witch and her blood had to be released to destroy her power.

Walton was elderly, a loner, and his behaviour was eccentric. The villagers believed he had power over animals and birds. He bred huge

---

5. This case and Case III-2 are from accounts that were compiled from police and criminal court records. This presents further problems of fact and meaning. The original records were compiled under the control of persons with particular social and legal conceptions and with their own professional interests, which will colour their interpretations of statements made by the participants.

toads, and he was said to harness them to miniature ploughs and set them loose in his neighbours' fields. Witches in the sixteenth and seventeenth centuries were accused of using the same technique to blight their neighbours' crops. Walton was related to a person burned as a witch in a nearby village centuries before.

The nature of the murder and the fact that Meon Hill has an ancient Iron Age fortification led the noted demonologist Margaret Murray to postulate that Walton was killed because he practised witchcraft. February 14 in 1945 was both St Valentine's Day and Ash Wednesday. It was the date upon which the ancient Druids made sacrifices to ensure plentiful crops, and 1944 had been a bad year for harvests and for livestock.

Despite intensive investigations by Detective Superintendent Robert Fabian ('Fabian of "the Yard"'), no one was charged with Walton's murder. It is said that the villagers were unhelpful in his investigations. Because of the apparently ritual nature of the crime, for years Fabian returned to Lower Quinton on the anniversary of the murder in the hope of discovering something. Each year he hid on Meon Hill, much to the amusement of some of the locals (Hough 1991: 37–43).

In each of these four cases, from four markedly different societies in widely separated parts of the world, someone is believed to have used a personal mystical power to cause death, or some other misfortune, to other persons. The culprit has done so out of malice, directed against individuals with whom she or he ought to have harmonious, and often close and friendly, personal relations – husband, neighbours, the wife of a 'sister's' son. In three of these examples it appears that malicious use of such power runs in families. (Consequently the child of Winnemucca's young wife is killed with its mother.) Most of the suspects probably were persons of a type commonly accused of witchcraft in many societies, such as the elderly, the eccentric, and the misanthropic.

Because the witch may not be the only unsocial user of mystical power in a culture, one of the earliest anthropological definitions of witchcraft involved an attempt to differentiate it from another personally-held power, sorcery, on the basis of the kinds of techniques employed by the practitioner. It was claimed that witchcraft is an innate power that is activated by the possessor's malicious thoughts. No techniques are necessary. Indeed, the possessor may be unaware that he or she has this power, as were Hagabulu and Sidawɔ, and it may be identified only by its effects. This belief has been widely reported by ethnographers, from areas as far apart as East Africa (Evans-Pritchard 1976; Harwood 1970; Middleton 1963), West Africa (Bohannan and Bohannan 1953; Jones 1970), and New Guinea (Schieffelin 1977). In some cultures it is believed that such a power has a physical presence in the body, which can be discovered by autopsy (Bohannan and Bohannan 1953; Harwood 1970; Wilson 1963). On the other hand, sorcery is a learned skill requiring deliberate action, such as the use of spells, charms and medicines.

The reason for the wide acceptance of this distinction was because some of the earliest and most influential anthropological analyses of witchcraft concerned peoples of East Africa, where this differentiation of mystical agents appears to be common. The most influential study was Evans-Pritchard's pioneering examination of witchcraft and magic among the Azande of the Sudan–Zaire border (Evans-Pritchard 1976). However, many peoples do not make this distinction. In fact cultures demonstrate a range of ideas about mystical powers, and a variety of ways of inflicting mystical harm. Some peoples, such as the Central African Azande (Evans-Pritchard 1976) and the East African Safwa (Harwood 1970), make a clear distinction between mystical harm which results from innate power and that which is due to magic. Others, such as the Navaho of Arizona and New Mexico (Kluckhohn 1967; Simmons 1980: 135–46), may recognize different kinds of practitioners of evil powers, but regard them all as variants of a common evil. Peoples believe in the operation of a number of kinds of mystical powers, but their nature may vary between cultures and different emphasis may be placed on their importance. It is not possible to give an exclusive definition of witchcraft and, correspondingly, of the witch. Because of this some authorities have chosen to group all types of personal mystical power used for antisocial purposes under the generic term 'witchcraft' (Douglas 1967; 1970). However, I suggest that we can identify the witch as a specific type of mystical evil-doer, and that there *is* a sense in which his or her witchcraft is *internal*. But the internal aspect is an attitude of the user, rather than the manner of operation of the power involved. In some cultures witches are believed to contain a mystical power within themselves. In many others the malevolent individual must use external techniques (for example, reciting spells) in order to work harm. But in the majority of cultures people believe that there exist persons who seek to harm others, by mystical means, out of envy or spite or hatred.

Witchcraft is associated with an antisocial disposition – with spite, envy and malice. Some mystical techniques may operate irrespective of the personal disposition of the user. If we choose to call the deliberate use of techniques 'sorcery', then when the sorcerer is held to have acted out of malice he or she has also become a witch. Consequently, in some societies the same practitioner may be regarded by some persons as a witch but not by others, or as a witch at some times but not at others. In witchcraft it is the nature of the person using the power that is given significance rather than the source of the power, although in Continental Europe, to the educated classes, the source of the power came to assume as much importance as the user's character. Because of the nature of witchcraft, usually it is regarded as unambiguously evil and its use as unjustified. Because of his or her antisocial disposition, the witch is set aside from the ordinary run of humanity. In many societies, witches, and the associations they form, are thought to embody the antithesis of proper human behaviour. They offend against all that is held in the highest moral value, and they provide an anti-model of human society (Lewis 1976: 81; Lewis 1986: 56; Mair 1969: 15–21).

In many societies there are persons who are believed to possess the same power as witches but are expected to use this power to *defend* the community, to punish transgressors within and enemies from without and to protect the community against witches (Harwood 1970: 67–76). In Europe they were termed 'white witches'. Other cultures may not use this symbolism, so anthropologists sometimes describe such persons as 'anti-witches'. Because they possess the same power as do witches, people who are regarded as defenders of the community in one context may come to be regarded as witches in another. Because identification as a witch depends upon a perceived quality in the suspect, people may be regarded as witches by some persons and not by others; and at some phases of their lives and not at others. This is exemplified in the case of Stürmlin that opened this chapter. She had been regarded as a practitioner of benevolent magic, but now at least some people were beginning to think that she was a witch. Whether or not the power some individuals are believed to possess is regarded as witchcraft, rather than some other ability, is essentially a subjective judgement by their fellows. All this demonstrates how misleading it is to attempt a rigid definition of witchcraft.[6]

I argue that it is not a power that constitutes witchcraft, but rather the uses to which it is perceived to be put. It is the intentions of the user of the power that serve to classify it as antisocial in any particular context, and often this is related to the disposition of the user. Since this condition may be difficult to define with any precision, in many cultures no precise line may be drawn between 'witches' and some other mystical practitioners.

In the great majority of societies, people view the world as governed by a set of personalized forces – in contrast, for example, to the depersonalized idiom used by the contemporary Western scientist, who sees the world as made up of atoms, molecules, and similar non-animate

---

6. On this point, see Briggs 1989: 15–16, 26, for French-speaking areas of Early Modern Europe.

entities, acting in a mechanical fashion. Explanatory theories are an essential aspect of all human cultures, as they are used to make the world appear ordered and predictable. In most societies, the entities used to create this order and predictability have usually been modelled upon the world of interpersonal relations. The world is controlled by personalized, extra-human powers that influence the lives of communities and individuals, and with which humans may enter into social relationships. Often they are concerned with the morality of human actions and events. The idea of the witch, that involves beliefs such as shape-changing, night flights to obscene meetings, cannibalism and animal familiars, is a part of this view of the world (Horton 1967).

Societies that have this idea of the witch are predominantly rural and agricultural, or have some other type of non-industrial economy, with much of life lived in small groups or communities whose members interact constantly and have relatively intimate relationships. Relationships are what are termed 'multiplex', the same individuals often occupying a number of different social roles with regard to each other. Relationships are multi-purpose, and are often kinship relationships (Gluckman 1959: 95; 1965: 256). It is because they share these characteristics that I believe I am justified in making deductions on witch-beliefs by comparing data from societies as divorced in space and time, and as different in certain aspects of their structure, as traditional Native American and African societies and the communities of Early Modern Europe. In all of these, the world-view held by their members is of the type defined. In all, their members live in small, enclosed worlds of intimate personal relationships. Social change takes place slowly, and the level of technology is low.

However I do not wish to imply that we can make a simple dichotomy between 'traditional' cultures with a non-'scientific' view of the world, and 'modern' cultures with a 'scientific' world-view. The scientific attitude to things and events accepts the fallibility of existing theories and is prepared to jettison them or reduce their importance when their efficiency as tools for explanation and prediction declines, and efficiency is measured through constant querying and experimentation. It is used consistently by relatively few persons even in our 'modern' industrial societies (Horton 1967: 185–6). (This is worth remembering when we come to consider the operation in 'modern' societies of mystical beliefs analogous to those concerning witches in other societies and cultures – for example, beliefs in contemporary Europe about 'coloured immigrants', and the scapegoating of racial and ethnic minorities.) Some persons hold a personalized world-view (as is demonstrated by the case of Charles Walton), and many, probably the majority, find a depersonalized world-view inappropriate in some situations and for some

A Deed without a Name

spheres of activity, because it is essentially amoral and reductionist.[7]

In societies with a personalized idiom of the world, witchcraft is often one of the ways in which people explain misfortunes, personal or social. Where it is invoked as an explanation for apparently undeserved misfortune (Lewis 1976: 73–5; Mair 1969: 9–10), it often seeks to explain what we would call the chance element of misfortune – why it should happen to one particular person or group and not another, and why a misfortune occurred when it did. Monica Wilson quotes a Pondo teacher in South Africa in the mid-twentieth century: 'It may be true that typhus is carried by lice, but who sent the infected louse? Why did it bite one man and not another?' (Wilson 1970: 262). The answer is that it was sent deliberately by the witch, to bite the man that it did.

In a classic passage Evans-Pritchard describes the explanatory role of witchcraft beliefs among the Azande of Central Africa as he studied them in the late 1920s (Evans-Pritchard 1976: 21–3). His description still stands as a general statement of the principles behind the operation of witchcraft beliefs.

> In speaking to Azande about witchcraft and observing their reactions to situations of misfortune it was obvious that they did not attempt to account for the existence of phenomena, or even the action of phenomena, by mystical causation alone. What they explained by witchcraft were the particular conditions in a chain of causation which related an individual to natural happenings in such a way that he sustained injury. The boy who knocked his foot against a stump of wood did not account for the stump by reference to witchcraft, nor did he suggest that whenever anybody knocks his foot against a stump it is necessarily due to witchcraft, nor yet again did he account for the cut by saying that it was caused by witchcraft, for he knew quite well that it was caused by the stump of wood. What he attributed to witchcraft was that on this particular occasion, when exercising his usual care, he struck his foot against a stump of wood, whereas on a hundred other occasions he did not do so, and that on this particular occasion the cut, which he expected to result from the knock, festered whereas he had had dozens of cuts which had not festered. Surely these peculiar conditions demand an explanation. Again, every year hundreds of Azande go and inspect their beer by night and they always take with them a handful of straw in order to illuminate the hut in which

---

7. For this reason, and because I believe Horton's view that science is an attitude rather than an idiom (Horton 1967: 69–70), I do not use Mary Douglas' dichotomy of world-views into Copernican and pre-Copernican, where a pre-Copernican world-view sees the universe as Man-centred, organized in relation to the lives of humans. It is expected to behave as though it were intelligent, responding to signs, symbols, gestures and gifts – for example, to prayers and sacrifices. Such a universe is considered able to make judgements about the moral value of human acts, and explanations of events are made in terms of the good or bad fortune or the good or bad behaviour of individuals and human groups (Douglas 1966: 80–8).

it is fermenting. Why then should this particular man on this single occasion have ignited the thatch of his hut? Again, my friend the wood-carver had made scores of bowls and stools without mishap and he knew all there was to know about the selection of wood, use of tools, and conditions of carving. His bowls and stools did not split like the products of craftsmen who were unskilled in their work, so why on rare occasions should his bowls and stools split when they did not split usually and when he had exercised all his usual knowledge and care? He knew the answer well enough and so, in his opinion, did his envious, back-biting neighbours. In the same way, a potter wants to know why his pots should break on an occasion when he uses the same material and technique as on other occasions; or rather he already knows, for the reason is known in advance, as it were. If the pots break it is due to witchcraft.

We shall give a false account of Zande philosophy if we say that they believe witchcraft to be the sole cause of phenomena. This proposition is not contained in Zande patterns of thought, which only assert that witchcraft brings a man into relation with events in such a way that he sustains injury.

In Zandeland sometimes an old granary collapses. There is nothing remarkable in this. Every Zande knows that termites eat the supports in course of time and that even the hardest woods decay after years of service. Now a granary is the summerhouse of a Zande homestead and people sit beneath it in the heat of the day and chat or play the African hole-game or work at some craft. Consequently it may happen that there are people sitting beneath the granary when it collapses and they are injured, for it is a heavy structure made of beams and clay and may be stored with eleusine as well. Now why should these particular people have been sitting under this particular granary at the particular moment when it collapsed? That it should collapse is easily intelligible, but why should it have collapsed at the particular moment when these particular people were sitting beneath it? Through years it might have collapsed, so why should it fall just when certain people sought its kindly shelter? We say that the granary collapsed because its supports were eaten away by termites. That is the cause that explains the collapse of the granary. We also say that people were sitting under it at the time because it was in the heat of the day and they thought that it would be a comfortable place to talk and work. This is the cause of people being under the granary at the time it collapsed. To our minds the only relationship between these two independently caused facts is their coincidence in time and space. We have no explanation of why the two chains of causation intersected at a certain time and in a certain place, for there is no interdependence between them.

Zande philosophy can supply the missing link. The Zande knows that the supports were undermined by termites and that people were sitting beneath the granary in order to escape the heat and glare of the sun. But he knows besides why these two events occurred at a precisely similar moment in time and space. It was due to the action of witchcraft. If there had been no witchcraft people would have been sitting under the granary and it would not have fallen on them, or it would have collapsed but the people would not have been sheltering under it at the time. Witchcraft explains the coincidence of these two happenings.

Witch-Beliefs and the Structure of Society

The importance of witch-beliefs as an explanation for misfortune varies in different societies, and of some it is recorded that they possess no belief in witches. The Mbuti pygmies of Zaire are one example, as recorded by Turnbull from fieldwork he carried out in the 1950s (Turnbull 1984).

The Mbuti are hunters and gatherers living in the Ituri rain forest in northwest Zaire. They live in small villages of fairly stable composition and have a symbiotic relationship with villages of Bantu cultivators with whom they exchange meat for horticultural products and who perform initiation rituals for Mbuti boys. The Bantu have a strong belief in witches. It appears from Turnbull's account that different Mbuti individuals have different attitudes towards witchcraft. Most Mbuti do not believe in it, but through their relationship with the Bantu a few have come to accept its existence, while others are uneasy about it. Turnbull describes the case of Aberi, a Mbuti man who was 'one of the few who really believed in it [i.e. witchcraft]' (Turnbull 1984: 205). However, from our point of view what is significant in this example is the attitude of Aberi's fellow Mbuti to his death.

Case II-6

Aberi and two other pygmy men made an agreement with a BaNgwana Bantu villager to give him a lot of meat in exchange for a sack of rice. The villager gave them the rice but never received the meat. The villager was believed to be a witch. He possessed witchcraft substance in his body, which he used deliberately to harm others. He cursed the three pygmies, and Aberi believed the curse would kill him. One night in the BaNgwana village, when he was going to the latrine, he was seized from behind by an invisible hand and thrown to the ground. From that time onwards he suffered from violent dysentery. One night in the Mbuti camp he awakened suddenly and struggled to get up as though an invisible hand was clutching his shoulder. Then he gave a loud cry and fell back dead. Within weeks the other two pygmies had died under similar circumstances.

Although they knew of the curse, Aberi's fellow Mbuti did not attribute his death to witchcraft. In so far as they mentioned the BaNgwana man it was only to postulate that he might have used poison to kill the three men – a speculation Turnbull clearly regarded as impossible (Turnbull 1984: 206–7).

A more common situation than that among the Mbuti appears to be where there is a belief in witches, but where witchcraft plays little part in peoples' lives and accusations are uncommon. Baxter discusses the situation among East African pastoralists where, although the idea of the witch is well developed, witchcraft is of minor importance in human affairs. Witchcraft accusations are rare, as is the attribution of witchcraft as the likely cause of illness in any specific case (Baxter 1972). Baxter contrasts the situation among pastoralists with that of many cultivators, where neighbours are bound together by property rights in land and often see themselves as competitors for common and limited resources. Rivals find it difficult to move away from one another without creating an absolute rift in their relationship. In contrast the nature of pastoralist property relations, in particular the diffuse nature of rights in livestock, and the relative ease with which persons can move apart in the normal course of pursuing a pastoral existence, means that tense, conflictual personal relationships do not have to be a major aspect of everyday life. Consequently, he argues, the cosmos does not give an important place to occult actions performed by malevolent individuals. Among East African pastoralists there is an inverse relationship between accusations of witchcraft and sorcery and the relative ease with which homesteads can move without creating economic or social disruption. Of Baxter's sample of nine pastoral societies, those with the highest reliance on agriculture in their domestic economy demonstrate the highest level of witchcraft or sorcery accusations (although the level is low for all nine societies).

However, witchcraft accusations are rare also among some agricultural peoples, including some with high population densities. The Ibo of southeastern Nigeria, some of whom have a population density of 1,000 persons to the square mile (Forde and Jones 1950: 10–11), believe in witches, but accusations are extremely rare. This is in marked contrast to some neighbouring peoples, such as the Efik, Ekoi, Mbembe, and Bekworra (Jones 1970: 325–7). G. I. Jones argues that this is due to the nature of Ibo social relations, which are less conflictual and allow conflict to be expressed more freely than do those of their neighbours. He argues that one reason for the relative lack of conflict is the Ibo pattern of territorial expansion. Typically, the Ibo local community is a group of villages, often related by ties of agnatic descent, so that the villages are seen as the local branches of a single patrilineage. The pattern of expansion did not bring land-owning units (hamlets and villages) of the same local community into conflict. When the expansion of individual communities ceased to be possible, as has been the case in the central Ibo area since at least the eighteenth century, many Ibo became migrant craftsmen, and this eased the pressure upon local land resources. Economic expansion and specialization ameliorated the effects of the limits to territorial expansion.

Some social anthropologists argue that certain kinds of social relationships give rise to witchcraft beliefs and accusations; that both cosmology and action are a function of social structure – of the pattern of social relations. They argue that the nature of social relationships in a society is the main determinant of the type of beliefs its members hold and of the way they behave towards each other. Consequently, one would be able to predict the presence or absence of witchcraft beliefs and accusations from a given pattern of relationships. In her Introduction to a symposium on *Witchcraft Confessions and Accusations*, Mary Douglas summarizes such social structuralist thought on witchcraft and sorcery when she argues that witchcraft accusations are most prominent where the pattern of relationships has two characteristics. The ideal component (the set of rights and obligations regarded as proper to specific relationships) is ambivalent or ill-defined, so that the content of one relationship is not clearly differentiated from that of another, and consequently relationships are conflictual; and the network of relationships is dense, with regular contact between individuals (Douglas 1970: xxx–xxxiii).

In another work, published a few years later (Douglas 1973: 136–52), Douglas again argues that witchcraft occurs in societies with two specific characteristics. Firstly, the community's control over the actions of its members, its ability to bring pressures to bear upon them, is strong (a condition she terms 'strong group'.) Secondly, with regard to the pattern of social relations within the community, statuses (i.e. social positions) are not differentiated clearly from each other, so that their associated roles – the ideal patterns of behaviour between persons – are not clearly defined and the rules of succession to offices and statuses are ambiguous (a condition she terms 'weak grid').[8] Such a situation promotes conflict between relationships and competition for offices and statuses. (An example would be the kinds of competition between aspirants for the office of village headman in some African communities that are discussed below in Chapters 6 and 7.)

Douglas argues that this pattern of relationships promotes the belief in sinister powers operated by fellow humans. For her, the typical witch-haunted society is the small community with a strong identity, but within which relationships are competitive. Witchcraft accusations are used to denigrate rivals and reduce them in the competition for leadership, and witchcraft beliefs attribute misfortunes to the moral failings of individuals and operate to uphold the moral law – to reinforce the system of social controls.

Douglas is perhaps the strongest exponent of structural determinism of beliefs. For her, the only way to change from a witch-dominated

---

8. The terms 'group' and 'grid' are Douglas' own coinages.

cosmology is to change the social structure – to change the pattern of social relationships (Douglas 1973: 146). However, quite apart from the difficulty of measuring degrees of 'group' and 'grid', her analysis would not appear to apply to activities in complex, contemporary, industrial societies that are analogous to witch accusations and persecutions in less complex societies, such as racism and McCarthyite political witch-hunts. It would appear that in these societies, although roles are competitive, the individual is relatively free of social control. Her explanation also fails to account for why men so often ascribe witchcraft to women.

Moreover, it appears that witchcraft and sorcery beliefs can be well developed among hunter–gatherers (Service 1966: 68–70, 88–110), who are often highly mobile, with ease of movement between bands enabling persons who develop antagonistic personal relationships to separate from each other. This was the situation among the Paiute, for example (Fowler and Liljeblad 1986; Steward 1938).[9] Accusations might be fairly common in such societies. (However, their frequency may be difficult to assess for the time prior to their subjection to the disruptive consequences of European colonial expansion, which may have had the effect of increasing the incidence of accusations.)

It appears that the pattern of beliefs in a society has a degree of independence of its social structure (Gluckman 1965: 226–7; Lewis 1976: 98–105), that it is not simply a reflection of its pattern of social relationships. Anthropologists have failed in their attempts to explain such inter-cultural differences in broad social terms. If we return to the case of the Ibo, for example, Jones' explanation in terms of degree of competition over land would appear inadequate as an explanation for the lack of variation in belief between different Ibo groups, or the great contrast in the social importance of witchcraft beliefs between the Ibo and their Cross River neighbours such as the Yakö and the Mbembe.[10] However, where witchcraft beliefs are found – and they are very widespread, presumably because they are a likely consequence of a personalized interpretation of the world – we can say that, at least during periods of relative stability, *the incidence of accusations* often appears to be influenced by the nature of relationships within a society.

Cultures that have the belief that individuals can obtain personal mystical power that can be used for the benefit of their fellows also have the belief that it may be misused, and persons believed to possess such power may provide the main source of witches in hunter–gatherer societies. In these societies witchcraft is the dark aspect of shamanism.

9. In Steward 1938, two of the subject's sons appear to have been killed as suspect witches.
10. Dr Rosemary Harris, personal communication.

Also, in many types of society certain categories of persons may come to symbolize values that are antithetical to the moral operation of social life. This is irrespective of the general pattern of relationships in the community. Such persons may be thought to be witches, even if they are rarely accused. From Baxter's sample of pastoral societies it is significant that the persons who are accused of witchcraft are regarded as being in some sense 'outsiders' to normal society, 'strangers' who do not really belong, such as persons of Turkana descent among the Samburu and Yibir bondsmen among the northern Somali (Baxter 1972: 165–9).

In the following six chapters I shall be examining data on witchcraft from African local communities in various periods of recent African history, from native American communities in the USA, and from England in the sixteenth and seventeenth centuries. Witchcraft beliefs and accusations have the same basic form in these societies. This differed in some significant respects from the classic European idea of witchcraft that developed in Continental Europe during late medieval times and is the subject of Chapters 9 and 10. From these examples, I hope to deduce some generalizations about the nature of witchcraft, and particularly about the kinds of people who stand in danger of being accused of its practice. The non-English examples are taken mainly, though by no means exclusively, from societies from east and southeast Africa and the southwestern area of the United States of America. These regions have produced a particularly rich and analytical literature on the operation of witch-beliefs. Indeed, they produced the two seminal anthropological monographs on witchcraft – Evans-Pritchard's *Witchcraft, Oracles and Magic among the Azande* (first published 1937), and Kluckhohn's *Navaho Witchcraft* (first published 1944). For England we have the detailed examinations by Keith Thomas (Thomas 1973) and Alan Macfarlane (Macfarlane 1970a). These examples are valuable for general comparison, as the three regions are far apart on the globe and there were many cultural, social, and economic differences between them.

Chapters 9 and 10 deal with the history of witchcraft in Europe. By the end of the Middle Ages, on much of the continent of Europe, witchcraft had become transformed legally into a heresy. Trials were no longer concerned with mystical harm, but with a pact made between the witch and the Devil. The very fact of being a witch had become a more important offence than the use of one's powers to harm others.

Medieval England was largely isolated from these Continental developments, owing to the independence of the English Church. England had no Inquisition or Roman Law, and papal authority was more limited. Causing harm by witchcraft continued to be the main concern of witchcraft trials. Legally, witchcraft remained an antisocial activity, rather than a heresy (Thomas 1973: 251–3, 527). English witchcraft, then, is of the general type discussed in Chapters 3 to 8, even though it came to be believed that the witch obtained her power from the Devil. English witchcraft cases rarely involved allegations of Devil-worship. Even rarer are references to a witches' sabbat, where witches reputedly met to dance and feast, indulge in sexual orgies and worship the Devil.

However, recent historical studies of the operation of witchcraft beliefs in post-medieval continental Europe are revealing that at the local level, in the village community, the pattern of witchcraft remained similar to that in England. (See, for example, Briggs 1989: 7–65, especially 31–7; Ladurie 1987a: 5–78, especially 59–62, 74–5.) Accusations of witchcraft were related to arguments and hostilities between neighbours, and the concern primarily was with misfortunes to persons and property, and not with pacts with the Devil and Satanic nocturnal orgies. It was only when accusations reached the courts that these elements were emphasized, by the clergy and the magistracy. Locally, misfortunes were remedied and witches identified by magical experts, who were condemned by the Church. Within the local community, concerns with witchcraft were treated as a local matter unless a suspect acquired such a bad reputation for witchcraft that she was regarded as a public menace. At the local level then, by the peasant and the labourer, the witch was regarded in the same way and treated in the same manner as in England. This pattern preceded the classic European conception of the witch, remained relatively unchanged during its heyday, and continued after its demise. As one authority says, it was 'made up . . . of a rural substance much anterior to the definitive late Medieval crystallization of the (demonic) stereotype' (Ladurie 1987a: 77). Consequently, when I discuss English witchcraft, sometimes I will also include data on rural witchcraft on the continent of Europe.

–3–

The Idea of the Witch

In many cultures the image of the witch is a truly horrific one. The witch is the personification of evil. Witches contravene all the basic values of society. They indulge in the worst kinds of antisocial actions, offending against the most deeply held and emotive mores of a culture. Consequently, often they are believed to indulge in forbidden sexual activities such as incest, intercourse with animals, and necrophilia. Their offences against kinship go beyond sexual deviance. Frequently they are believed to kill their own kinsmen – those persons to whom they have the strongest obligations – to enable them to achieve their own selfish material ends. Witches may indulge in cannibalism, real or mystical, of their victims. Thus, not only do they kill, they also consume their kin and the kin of their fellow-witches, similarly murdered. Their actions are governed by hatred and envy to such a degree that they will kill or cause harm simply out of malice and often in a random fashion. They are intimately associated with that which a culture regards as evil or as suspect in some way. They operate at night, often in close association with animals identified with night and with evil. They may have such animals as familiars. They may be associated with that part of the geographical world which is regarded as dangerous, ambivalent, or evil. For example, they may carry out their nefarious activities in the wild 'bush' rather than in the settled, and social, community (Mair 1969: 38–45).

To give an example, the Shona witch is always jealous, is aware that she is a witch, and enjoys her evil behaviour. She is driven relentlessly to kill. She may cause harm for no reason and may attack members of her own family. The most powerful Shona witches always operate at night. They have animals such as owls and snakes as familiars and send them to bewitch their victims. A witch rides naked to her nightly meeting with her fellows on a hyena, an ant-bear, an otter or a pig. Witches exhume the bodies of their victims and eat their flesh, and they use human flesh to make their most powerful 'medicines' (Crawford 1967: 111–27; Gelfand 1967: 11–31).

Kluckhohn provides much information on the Navaho idea of the witch (Kluckhohn 1967: 22–34). The Navaho occupy a large reservation in the

'Four Corners' area of the southwestern United States (the junction of Arizona, New Mexico, Colorado, and Utah) (Ortiz 1983: 489–683). In 1970 they numbered approximately 130,000. Most of their arid 25,000 square mile reservation is unsuited for agriculture. Traditionally they are sheep-herders and farmers, living in small scattered neighbourhood groupings. In rural locations they generally continue to live in extended families – small clusters of homesteads containing the families of a mother and her daughters. Today there is some development of urbanization, and industrial and commercial developments on the reservation are having an increasing effect on Navaho life. There is also seasonal emigration for work off the reservation, such as work in the beet fields of southern Colorado. Since 1938 there has been a Navaho Tribal Council, operating under the supervision of the Bureau of Indian affairs.

These changes have produced strains and conflicts in Navaho society. One response to this has been the growth of membership of the Native American Church, whose ritual is organized around the taking of peyote, a hallucinogenic cactus, and which blends traditional Navaho ceremonialism with fundamentalist Christian elements and 'pan-Indian moral principles' (Dutton 1975: 34). Traditional Navaho religion centres on curing ceremonies performed by singers and curers, intended to restore health, secure food, and ensure survival. The Navaho have a great fear of ghosts and witches (Dutton 1975: 6–36).

Kluckhohn's examination of Navaho witchcraft is concerned particularly with the 1930s, when economic depression in the USA intensified the influence of American society upon the Navaho, increasing social tensions and producing a resurgence in witchcraft fears and in assaults upon persons suspected of witchcraft (Kluckhohn 1967: 114–17). Among the Navaho, witchcraft is associated with death, the dead, incest, the night, and certain animals such as the coyote and the owl. Witches attack their victims with a powder made from human corpses, preferably the corpses of children, and particularly of twins. At night they may assume animal forms, often those of wolves or coyotes, and operate as were-animals. Navaho believe witchcraft usually is learned from a parent or grandparent or a spouse, and that initiation into witchcraft involves the initiate's killing a close relative, usually a sibling. They believe that people become witches in order to pursue selfish or antisocial personal and material ends. They desire vengeance against someone or wish to acquire wealth; or they may simply wish to kill wantonly. Wealth is obtained by robbing graves or by fee-splitting. A witch enters a partnership with a ceremonial curer. The witch makes someone ill and his partner receives a fee for curing the victim (Kluckhohn 1967: 22–34).

The Kaluli witch's evil manifestation, his 'witch child' (Schieffelin

1977: 101), creeps around at night in search of victims. Those with the ability to see it perceive a crouching, creeping figure with glowing red eyes, clutching hands, and a distorted face. It is the incarnation of evil – malicious, irrational, voracious, disgusting, dangerously violent and tremendously strong (Schieffelin 1977: 101–2). Acting on the mystical plane, it injures or dismembers its victims' bodies until finally it kills them by pulling out their hearts. Later it sits, invisible, on the edge of the coffin and eats the corpse, causing the flesh to disappear (Schieffelin 1977: 127). It is driven by a dangerous, unreasoning wilfulness or hunger. It should attack only those to whom its possessor is unrelated by blood or marriage; but when it is very hungry, or very angry, it attacks indiscriminately. Where the reasons for an attack are known, they are trivial – a minor theft or insult or an imagined slight.

Among the Kaluli exchanges of food are the medium through which people express their social relationships. Anyone who does not deal properly with food is said to behave like a witch (Schieffelin 1977: 46). Nearly all witches are men, and their witch-aspect emerges at night, from the penis. Kaluli witchcraft is a secret form of male potency that operates for evil purposes. Consequently, when a witch is killed his genitals are cut off (Schieffelin 1977: 101–2, 127).

Among another New Guinea people, the Tangu, who live inland from Bogia Bay in northern New Guinea, the concept of the witch comes within the Tangu concept of the *ranguma*. Tangu society is based upon a principle of equivalence, again expressed in food exchanges. The *ranguma* is always a man, one who does not behave in a reciprocal way. He takes, rather than exchanges. The term covers a wide variety of deviance and in different contexts it can be translated as 'sorcerer', 'witch', 'criminal', 'assassin', 'scapegoat', 'non-conformist', or even simply as 'unusual' (Burridge 1965: 230). Tangu say that the behaviour of the *ranguma* is *imbatekas*, a term they apply to any act that is not considered normal, preferred, expected or commonplace (Burridge 1965: 230). This includes acts which could be considered 'evil'. Essentially they are acts considered non-reciprocal or uncontrollable.

The image of the *ranguma* conforms to the common witch-image. He is thought of as a tall, bony man with red-rimmed eyes and the splayed hand and long fingers of a strangler. He is surly and unsociable. He has abilities not possessed by ordinary men, but he uses them for his own selfish ends. He may sell his services to perform harm for others. He deliberately incites immoral behaviour in others, intentionally causes persistent illness and initiates killing outside of warfare and feuding. He kills with the precision of a professional, using a variety of means (mystical and non-mystical) in order to do so, and he is particularly active at night.

The Idea of the Witch

The idea of familiar spirits was well-developed in the English witchcraft of post-medieval times (Macfarlane 1970a: 6; Thomas 1973: 527–9). Familiars are part of the English folk-concept of witchcraft. In the sixteenth century, clerics incorporated them into the concept of Satanic witchcraft by arguing that the Devil gave to witches personal spirits, with whose assistance they carried out their nefarious actions (Holmes 1993: 65–75; Macfarlane 1970a: xxi). These usually took an animal form, often of a domestic type. In Essex, they frequently took the form of a cat or a toad (Macfarlane 1970a: xxi). They were given names that demonstrated the personal nature of the relationship between witch and familiar.

The witch suckled her familiar on her blood through some place on her body, which left a mark known as 'the witch's mark'. Interrogators looked for this as proof that a suspect was a witch. It was easy for interrogators to find a witch's mark. A supernumerary nipple or a wart might be taken as one, but a discoloration or a sore could do. And if a mark was absent, it was known that a witch might remove it or it might come and go (Peel and Southern 1969: 23; Thomas 1973: 657). However, the evidence suggests that women designated by the courts to search female suspects for incriminating marks were more likely to find them when the community believed the suspect guilty. In other words, searchers often embodied the opinions of the local community, and this guided their interpretation of any physical manifestations that they discovered (Holmes 1993: 74). In the legal prosecution of witchcraft in England, 'the witch's mark' replaced 'the Devil's mark' of Continental witchcraft. This was the mark the Devil put on the witch's body to seal their bargain (Holmes 1993: 65–75). It was part of the Continental European complex of witch beliefs that developed at the end of medieval times. (Its development is described in Chapter 9.) The Devil's mark was insensible to pain and, on a woman, its most likely location was the breasts or the pubic area.

English witches might be charged with a variety of offences. The great majority involved harm to animals, dairy products, and human beings. They would prevent cows from giving milk, interfere with the process of turning milk into butter and cheese, and make ale sour. Witches were most likely to be indicted for more serious offences such as the death or illness of humans and harm to farm animals, but they were also accused of other offences, for example, damage to farm carts and other inanimate property. Macfarlane states that in Essex the largest single category of offences involved human deaths. He estimates that they made up about 40 per cent of accusations, and they formed a larger proportion of indictments. Damage to property believed caused by witchcraft could be considerable. In 1593 a woman was accused of killing twenty-two sheep, one cow, one pig and one calf, with a total value estimated at £7 and 16*s*.

Another was accused of killing four geldings and sixteen cows, estimated value £62 (Macfarlane 1970a: 152–4). Witches could be blamed for a variety of illnesses, such as rheumatism, arthritis, and stroke. No specific illness or disease was always blamed on witchcraft, although strange, unidentifiable and inexplicable diseases were particularly likely to be attributed to witches. Strokes and epilepsy, and unusual illnesses in animals, for example (Macfarlane 1970a: 81–2, 88–9; Thomas 1973: 519–20, 540–3, 667–8). After 1660, English indictments of witchcraft focused upon the illness and death of persons rather than destruction of livestock and other property. However, in popular belief attacks upon livestock and interference with domestic and agricultural processes and procedures continued to be just as prominent in suspicions of witchcraft (Holmes 1993: 47–9).

On the other hand, plague was not associated with witchcraft, and there were other misfortunes that were unlikely to be blamed upon witches. They were rarely blamed for business failure, except by members of the upper classes who might seek to explain and denigrate a rival's success by attributing it to witchcraft. In contrast to what happened in continental Europe, English witches were not usually blamed for storms or for sexual impotence (Briggs 1989: 27; Macfarlane 1970a: 152–4, 178–83; Thomas 1973: 639–45).

In the French-speaking regions of post-medieval Europe, sudden late frosts and impotence frequently were attributed to witchcraft. (Impotence was believed to be caused by a witch tying a knot in a lace at the moment of marriage.) Although in theory almost any misfortune could be ascribed to witchcraft, people in these regions tended to be more discriminating. It was most likely to be blamed for illnesses in people and animals that were not readily identifiable or failed to respond to treatment, and sudden deaths of previously healthy individuals. Mental illness frequently was ascribed to witchcraft, and witches might be blamed for epidemics in animals and for failure of milk to churn or cows to give milk (Briggs 1989: 29–30; Ladurie 1987a: 25, 35–7, 40–4, 55, 64–6, 69, 70–5).

One of the most famous English witch trials, that of the Pendle witches in 1612, demonstrates many of the characteristics of English witchcraft. I cite it in some detail because it will be referred to again in the following chapters. (This account is taken from Peel and Southern 1969.)

Case III-1

The case involved two families of witches, the Redfearns or Chattoxes, and the Devices or Demdikes, who lived in that part of northeastern Lancashire known as Pendle Forest, an agricultural region that once had been set aside for hunting. The Redfearn family consisted of Anne Whittle, a widow, aged about 80 in 1612, known as 'Chattox'; her daughter Anne, with her husband Thomas Redfearn; and one or two other family members. At her trial Chattox was described as '. . . a very old withered spent and decrepid creature, her sight almost gone . . . her lips ever chattering and working but no man knew what' (Peel and Southern 1969: 58). The Demdikes lived a few miles away in a dwelling called Malkin Tower. Elizabeth Southern, 'Old Demdike', was a widow of the same age and poor economic position as Chattox. She was blind and lame. With her lived her widowed daughter, Elizabeth Device (Davies?), probably in her late forties and very ugly, and her children. Of these, James, her oldest child, possibly was mentally subnormal (Peel and Southern 1969: 43); Alizon was probably in her late teens; and Jennet was aged nine. The family lived mainly by begging.

On 18 March 1612 Alizon Device met a pedlar, John Law of Halifax. She asked him for some pins, and became angry when he refused. Almost immediately, he suffered a stroke. He was carried into an alehouse, crippled and unable to talk. When he recovered his speech he accused Alizon of bewitching him. His son, Abraham Law, brought her to his father and she confessed to bewitching him and begged his pardon. The father accepted her pardon, but the son laid a complaint of witchcraft before John Nowell, the magistrate.

Examined by Nowell on 30 March, Alizon confessed that she had crippled James Law with the help of her familiar spirit, a black dog. She said she had been initiated into witchcraft by her grandmother, Old Demdike, and accused her of being a witch and killing the little daughter of a local miller, Richard Baldwin. She said that two years earlier she led her blind grandmother to Baldwin's house to ask for payment for work that had been performed for him by Alizon's mother, Elizabeth Device. He threatened them, saying 'Get out of my ground, whores and witches. I will burn one of you and hang the other.' Demdike retorted, 'I care not for thee. Hang thyself.' Shortly afterwards the miller's daughter took ill and died.

Alizon then accused Chattox of killing her father, James Device. Device was afraid of Chattox, and gave her a yearly payment of meal on the understanding that she would not hurt any member of his family. The first year he failed to make the payment, he died. On his deathbed he said Chattox had killed him. Alizon also accused Chattox of killing a little girl for laughing at her. Chattox had warned the girl that she would get even with her. She had also killed the child of a man who accused her of turning his ale sour. Again, she warned the man that she would get even with him. She had killed Hugh Moore, a farmer who accused her of bewitching one of his cows. She had also killed the cow of a farmer, John Nutter, because his son deliberately kicked over a can of milk his father had given to her daughter, Anne Redfearn, alleging that she was using it to perform witchcraft. Alizon's brother, James, stated that Alizon had confessed to bewitching the child of one Henry Bulcock. Nowell had Alizon detained.

Three days later, on 2 April, Nowell questioned Demdike, Chattox, and Chattox' daughter Anne Redfearn. Demdike admitted she had been a witch for the past twenty years. (Others said it was fifty years.) She told how, twenty years before, she met the Devil in the form of a boy wearing a coat that was half black and half brown. He said his name was Tibb, and in return for her soul he promised to give her anything she desired. Tibb would come to her in the form of a brown dog, a black cat, or a hare, and suck blood from under her left arm, where she had a witch's mark. She admitted permitting Tibb to kill Richard Baldwin's daughter, and accused Chattox and her daughter Anne of killing Robert Nutter, the son of a local landowner, by image magic.[1]

Robert Nutter was the son of Christopher Nutter, on whose land the Chattoxes lived. In 1583 he threatened Thomas Redfearn, Chattox' son-in-law, that he would have his father put him out of his house. Redfearn is reported to have replied, 'When you come back again you will be of a better mind.' A few days later Nutter fell ill and died, having told his father that he believed he was bewitched by Chattox and Anne Redfearn. Shortly afterwards his father also fell ill. He told his family that he was bewitched, but named no one. After three months he died. These were the earliest deaths of which the Pendle witches were accused.

Chattox confessed to giving the Devil her soul ten years previously. He had appeared to her in a form 'like a Christian man' and promised she would want for nothing, and that he would enable her to revenge herself on anyone she wished. She was given a familiar spirit called 'Francie'. She admitted playing a part in the deaths of Robert and Christopher Nutter, about seventeen years before, but claimed that two other female witches were also involved, and they were more guilty than she. Robert Nutter's grandmother had asked the three of them to kill him, so that Christopher Nutter's land would be inherited by Robert's cousins. Chattox had been dissuaded by Thomas Redfearn. But then Robert Nutter tried to seduce her daughter, Anne Redfearn. Anne resisted, so Robert threatened to put them off the land. Chattox then called upon Francie for help to kill him. She also admitted the offences of which Alizon Device had accused her on 30 March, but she sought to play down their magnitude by saying that it was cows she had killed, not people, and that her actions were justified because she had been treated unjustly by those whose property she attacked.

Neighbours gave evidence that Chattox and Anne Redfearn were

---

1. Mutilating a clay image made in the form of the victim.

reputed witches. They had killed Robert and Christopher Nutter and committed other offences. Nowell detained Chattox, Demdike, and Anne Redfearn. Anne never made any confession. She was to be convicted on the confessions and 'evidence' of others.

On Good Friday, 10 April, children and friends of the accused held a meeting at Malkin Tower. It is said to have been a meeting of all the witches of Pendle Forest, and often is claimed to be one of the few examples of a witches' 'sabbat' – a congregation of witches to worship the Devil – in the history of English witchcraft. In fact it appears to have been a meeting of local persons who were afraid of being named by those arrested, and at the time no suggestion was made that the Devil had been present at Malkin Tower (Kitteredge 1958: 266). It was claimed that they plotted the release of the four prisoners in Lancaster Castle, and planned some witchcraft.

News of the affair quickly reached Roger Nowell, who carried out an inquiry. He questioned Elizabeth Device, her son James and her daughter Jennet. The latter two, a young man who may have been half-witted and a girl of nine, but particularly Jennet, became the principal witnesses against the other accused. They identified the persons who attended the Good Friday meeting and said that all were witches. James said he had a familiar spirit, a dog called Dandie, with whose help he had killed a Mistress Townley by image magic because she had accused his mother and himself of stealing. When John Duckworth promised him an old shirt and then refused to give it to him, he killed him with Dandie's help. He accused his mother, Elizabeth Device, of killing a farmer, John Robinson, by image magic.

Confronted with the accusations of her children, Elizabeth Device admitted the murder of John Robinson, saying it was because he accused her of having a bastard child. She also killed his brother, with the aid of her familiar spirit who was called Ball. She confirmed James' list of persons attending the Good Friday meeting. On the basis of this evidence, seven more persons, including Elizabeth and James Device, were sent to Lancaster Castle to await trial at the Assizes. Old Demdike died there of gaol fever. Chattox then claimed Demdike had been responsible for introducing her to witchcraft, and several other accused tried to implicate the dead woman in witchcraft in order to divert accusations from themselves.

Before the Pendle witches were brought to trial, Jennet Preston, one of their neighbours who had attended the meeting at Malkin Tower, was tried at York for witchcraft and hanged on 29 July. Although she was an associate of the Pendle witches, she lived in Yorkshire and consequently came under a different regional jurisdiction. She was a poor woman. Earlier the same year she had been acquitted on a charge of murder by witchcraft brought against her by the Lister family. This was a well-to-do family with whom she had once been on good terms, but who had come to blame her for losses of goods and cattle. It was claimed that one of the items discussed at the Malkin Tower meeting was help for her feud with the Listers. She was apprehended and charged with the murder, five years previously, of Thomas Lister. She pleaded not guilty and made no confession. Witnesses said that before he died Thomas named her as the witch who was killing him. The most damning evidence was the testimony that Lister's corpse bled when she touched it. James Device's account of the Good Friday meeting was read to the court. The jury found her guilty, and she was hanged two days later. It was claimed that she had a white foal as a familiar spirit.

The trials of the Pendle witches began on 18 August 1612. They were tried individually, before Sir James Altham and Sir Edward Bromley. Alongside them were tried three suspect witches from Salmesbury, making a total of twenty-three Lancashire witches tried at this Assizes. All but one pleaded not guilty.

All the Pendle witches were from poor families, with the exception of Alice Nutter, a gentlewoman who attended the Malkin Tower meeting and about whom there is little information, and Katherine Hewitt, known as 'Mouldheels', the wife of a clothier. Chattox pleaded not guilty, but was convicted on her own confession and that of Demdike. She tearfully acknowledged her confession, saying that her familiar, Francie, had failed to keep his part of their bargain. He had deserted her and taken away her sight. Her daughter, Anne Device, was convicted, partly on the evidence of her son James and her daughter Jennet. Nine-year-old Jennet testified against her brother, James. On the testimony of the dead Demdike, given on 2 April, Anne Redfearn was convicted of the murder of Christopher Nutter in 1595. Pedlar John Law's testimony convicted Alizon Device. He told how, as he lay paralysed in the alehouse, a huge black dog came and stared him in the face, just before Alizon came in to look at him.

Twelve of the suspected witches were found guilty, and ten were condemned to death. These included Chattox, Anne Redfearn, Elizabeth Device, James Device and Alizon Device. Margaret Pearson was found guilty of killing a mare. She had been tried and acquitted of witchcraft twice before. This time she was found guilty, and was condemned to be exposed in the pillory and sentenced to prison for one year. Only the poorest and mentally weakest of the accused confessed – Demdike, Chattox, Alizon Device, James Device and Elizabeth Device – and Elizabeth Device subsequently withdrew her confession. The three Salmesbury witches were acquitted when the chief witness against them admitted that she was lying. (She was the fourteen-years-old granddaughter of one of the accused, and the niece of another.) The three

The Idea of the Witch

accused had renounced their Roman Catholic faith, and the court regarded the accusations as part of a Catholic plot against them.[2]

On 20 August those condemned to death were taken from Lancaster Castle and hanged.

Twenty-one years later Jennet Device, then a woman of about thirty, was tried on a charge of witchcraft.[3]

Although the image of the witch often exhibits many common features between cultures, there may be significant differences in specifics. In a 1951 paper (see Wilson 1970), Monica Wilson contrasted the witch-beliefs of two African peoples among whom she had carried out fieldwork, the Nyakyusa of Tanzania (then the colony of Tanganyika) and the Pondo of South Africa. By the time of her fieldwork, in the 1930s, both had been incorporated into the colonial economy as migrant wage labourers, the Nyakyusa in the mines and on the sisal plantations of Tanganyika and the Pondo in the gold mines of South Africa. The Nyakyusa were also peasant producers of rice and coffee as commercial crops. Both were cattle-keepers and cultivators, and cattle were socially important as bridewealth for marriage. Among both peoples descent was patrilineal and there was an elaborate ancestor cult. Patterns of residence differed markedly, however. The Pondo residential unit was a homestead with a core of men of close agnatic descent together with their wives and children, with senior kinsmen having legitimate mystical power over their descendants and their junior relatives. In former times its members were held jointly responsible for civil wrongs committed by any member, and they fought as a unit in the district army. The Nyakyusa pattern was unlike that of most African peoples in that they lived in villages composed not of kinsmen, but of men of the same age. A wife lived in her husband's village. The Nyakyusa village was a land-holding unit. In the past its members made joint defence against enemies, and they jointly defended their fellows against witches.

Nyakyusa witch behaviour focused on food and milk. Witchcraft took the form of pythons in the bellies of individuals – witches. Pythons could be revealed by autopsy, and were inherited. Lust for meat and milk drove

---

2. The accusations in the case of the Salmesbury witches were much closer to the Continental idea of the witches' sabbat. They were charged with killing and eating a baby and rendering down its bones to make an ointment to be used for shape-changing, and with attending a meeting where they had intercourse with spirits (Peel and Southern 1969: 82–6).

3. In the 1633 case, although popular opinion was against those accused, the authorities were unconvinced and ordered a stay of execution. The seventeen convicted persons were reprieved when the chief witness, a boy of ten, admitted that his evidence was a fabrication (Holmes 1993: 66; Robbins 1959: 298; Thomas, 1973: 544 n. 69, 645).

such individuals to commit witchcraft. They flew by night, on their pythons or on the wind, and attacked singly or in covens. They delighted in eating human flesh, gnawing upon the insides of their victims until they died. Good men of the village mystically defended their neighbours against attacks by witches. Together they fought them in dreams. The source of their power was similar to that of witches themselves; they also possessed pythons in their bellies.

Among the Nyakyusa food and milk expressed and symbolized ideal relations within the village. Men should eat with age-mates of their own standing, and generosity with food and hospitality to neighbours gave prestige. In their lust for food and milk, and in their antisocial ways of attaining it, witches perverted the values and esteemed relationships of the group.

Among the Pondo the emphasis in witch belief was not on eating but upon sexual relations. The Pondo witch had a familiar – a hairy being with exaggerated sexual characteristics, or a baboon, a wildcat, a snake, or a lightning bird. The familiar was of opposite sex to its witch and often took human form, appearing as a very beautiful, light-coloured man or woman who had sexual relations with the witch. Familiars were totally evil and never used for purposes that basically were social in intent, such as punishing wrongdoers. Defending the homestead against witches was the responsibility of the ancestors or was achieved through use of medicines.

Wilson relates this difference of emphasis in beliefs about witch behaviour to the structure of the two societies, in particular to patterns of residence and to sexual regulations. The Nyakyusa lived in age villages, and Wilson interprets the value placed on hospitality between neighbours as a consequence of this. It was symbolized by eating and drinking the products of cattle, which were inherited by kinsmen who were scattered among different-age-villages. The Pondo, in contrast, were organized into large exogamous clans. As a consequence, many of the persons of opposite sex with whom a Pondo interacted were forbidden as sexual partners, and there was a strong aversion to clan incest. In addition, the Pondo were much more closely incorporated into a colour-caste society (White-dominated South Africa) than were the Nyakyusa. Wilson argued that the light colour of the familiar symbolized the forbidden sex attraction between members of different castes.[4] The emphasis on sexual relations in Pondo witchcraft beliefs she interpreted as the consequence of creating

---

4. Such psychological interpretations are difficult to substantiate. See Devons and Gluckman 1964: 241–53 for a concise criticism of psychological explanations of witchcraft made by non-psychologist anthropologists.

large categories of persons who are sexually forbidden to one another and yet are in close contact.

Wilson graphically characterized witch-beliefs as 'the standardized nightmare of a group' (Wilson 1970: 263). By 'nightmares' she means fears of attacks upon the group's deeply held values, and because some of these values differ between cultures the 'nightmares' are standardized in different forms in different cultures. Wilson believes that value differences are a consequence of differences in the structure of societies. The pattern of witchcraft accusations also differs between the Nyakyusa and the Pondo, and Wilson, like many others, interprets this also as a consequence of differences in structure. Nyakyusa mainly accuse neighbours but very rarely accuse kinsmen, with whom they interact less frequently. Pondo accusations are particularly common between a mother and her daughter-in-law, which is often a relationship of considerable friction. (We will be examining the relationship between structure and accusations in Chapter 6.)

The Witches' Society

We have seen that the witch is often the epitome of evil. Victor Turner relates this to the nature of many of the misfortunes attributed to witchcraft. In communities that do not have access to industrial technology many misfortunes, such as illness, are unpredictable and have a random and apparently motiveless character. The human perpetrators are conceived to have a similar character, being motivated by malevolence, often of a random kind, and a hatred for the moral order (Turner 1964: 315). Other persons have pointed out that if individuals are to be held responsible for causing misfortunes that are considered to be undeserved, then they have to be perceived as immoral individuals.[5] The implication here is that witch-beliefs have been developed in order to account for these misfortunes.

T. O. Beidelman on the other hand argues that the Kaguru witch is an imaginative product of the agonistic aspects of social interdependence. (The imagination is a 'moral imagination' because it makes judgements about what the world is and should be (Beidelman 1986).) It is a projection of the suppressed negative emotions that arise from being a member of society whilst at the same time being an individual – from the conflict that results from a person's occupying simultaneously these two, often mutually conflicting, positions. Consequently, although the witch is viewed as totally evil, Kaguru actions suggest ambivalence, rather than total rejection, towards the idea of witchcraft. Some attributes of the witch

5. Dr Rosemary Harris, personal communication.

are secretly coveted, such as power and aggressiveness. Others are strongly rejected, such as incest and cannibalism (Beidelman 1986: 156–7). The witch is an attempt to imagine beings that are morally outside society, that have rejected its moral constraints whilst continuing to reside within it. They have rejected their obligations to neighbours and kin, simply exploiting these relationships to achieve their own selfish ends. The result is a human whose humanity is denied, a monstrous being that behaves in inhuman ways (Beidelman 1986: 138–57). I find Beidelman's explanation appealing, as, in many societies, although the witch is perceived as totally evil, there is some ambivalence towards witchcraft and towards individual witches. However, whether or not Kaguru witches express the tension between individual and society, they are used to explain misfortunes considered undeserved.

The image of the witch shows many common characteristics worldwide. Witches commit incest. They kill their close relatives. They indulge in cannibalism. They have familiars and are identified with other creatures and situations that are regarded as evil, or anomalous. Witches oppose the moral order. Everywhere the basic moral attributes of the social order are the same – co-operation; respect for life; concepts of property; sexual regulations. The witch strikes at all these basic moral imperatives. His or her behaviour is a wilful attack on all that is considered most important. In many societies with a personalized world-view much that is basic to social life is bound up in concepts of kinship (Mair 1969: 36–42).

In many cultures witches are believed to act together in order to carry out their nefarious activities. Witch-life is ordered, and very often this order involves organized behaviour between witches. Often they have some form of organization in which many of the witches of an area, or all the witches of a culture, participate. Among the Shona, witches meet together at night, naked. Their favourite meeting place is the grave of someone recently dead, where they eat pieces of the deceased's flesh (Crawford 1967: 121–2; Gelfand 1967: 27–8). Frequently two or more Shona witches work together to bewitch their victims. Among the Navaho, witches meet together at night as were-animals to initiate new witches, commit acts of necrophilia and cannibalism, and carry out mystical killings. Their meetings are highly organized. The proceedings are directed by a Chief Witch who has two categories of helpers, senior and menial. In their daily lives the Chief Witch and his senior helpers are rich persons, whereas menial helpers are poor people (Kluckhohn 1967: 27).

Case III-2

In 1960 a young Shona girl, Shani, was found dead with a fractured skull. A circular piece of skin had been taken from her left cheek and some skin had been removed from her genitals. A woman, Ndawu, was charged with her murder. In her evidence to the preparatory hearing, 25 March 1960, Shani's mother, Muhlavo, testified that before she reached puberty she (Muhlavo) was instructed in midwifery and sex matters by a woman, Chirunga Tsatswani, who also initiated her into witchcraft. After she married, she and Chirunga Tsatswani went about at night bewitching people. One day a young girl, Ndawu, the daughter of her husband's second wife, came to her hut saying she wished to be friendly, and Muhlavo initiated her into witchcraft using medicine Chirunga Tsatswani had given her some years previously. Ndawu joined Muhlavo and Chirunga Tsatswani in their night-time activities.

One night Chirunga Tsatswani and Ndawu came to Muhlavo's hut, riding hyenas. They all went to the hut of Muhlavo's husband, Chidava, as they had dreamed that they must bewitch him. (Muhlavo claimed they had no other reason for harming him.) They poured some sweet beer containing bewitching medicine into his mouth, and sprinkled medicine on his body. Three days later he died. A few nights later the three witches exhumed Chidava's body. They skinned it, took a piece of meat, re-interred the body and went to Muhlavo's hut, where they cooked the meat and ate it. (As a result of Muhlavo's statement the police had Chidava's body exhumed, but found no evidence of its having been interfered with (Crawford 1967: 113).)

Some time later the three rode their hyenas to the kraal of Chidava's brother, Meke, and decided to kill him. They laid hands on the sleeping man and next day he was ill. The head of Meke's kraal came to them and asked them not to bewitch Meke, so they relented and he lived. After this Ndawu married and went to live elsewhere. Muhlavo said that one night she rode her hyena to visit Ndawu. Ndawu was holding her infant and Muhlavo tried to take it from her to bewitch the child so that it would die and they could eat it, an idea that came to her in a dream. Ndawu resisted and Muhlavo did not succeed.

Four days later Ndawu brought the child to Muhlavo and said it was sick and would not suck. Muhlavo went to the head of her kraal, who held a meeting of the kraal which decided that Ndawu should stay with Muhlavo. (Presumably this is confirmation that they believed Muhlavo had bewitched the child.) The following night the child died. Muhlavo took one of her children and went to report the death to Chief Neshuro, leaving Shani and her other child with Ndawu. When she returned Shani was dead. Muhlavo told the court she did not believe Ndawu had killed her child because she would have bewitched it, not killed it with a stick 'like a beast'.

Ndawu testified that she deliberately killed Shani by striking her with a pole, because Muhlavo had killed her child by witchcraft. Chirunga

Tsatswani testified that she and Muhlavo and Ndawu were witches and had killed Chidava by witchcraft and eaten part of his body. They had also bewitched Meke.

On 2 August 1960 the High Court found Ndawu guilty of Shani's homicide, but with extenuating circumstances, and she was sentenced to two years' imprisonment (Crawford 1967: 45–9; Gelfand 1967: 33–5).

Kluckhohn records the following story from a Navaho informant (Kluckhohn 1967: 146–7).

Case III-3

'A young boy was caught by two witch people. They said, "There is a sing[6] down here. We want you to go along." He thought a sing was going on over there where they said, so he made up his mind to go with them. It was after dark when he was caught. They start to going off in one direction. Kept going. I don't know how long they kept going. While they going they come to a place where there is some kind of a door in the ground. And the boy noticed the door was wide open. And he see the light out of that place. He looked down in there and seen a lot of people down in there.

When they come to that place the two witch people blow the same kind of whistle that the singer uses. After they come to that place, they went in there. After they got in there these other people they made a lot of talk about the young boy. They said, "Why did you bring this boy over. He's got no business to come in there." These people thought that the other people were bringing some dead people when they heard the whistle. That's why they open the door for them.

While they was waiting in there talking they heard a whistle outside. One jumped up and opened the door. Two people came in from the top carrying a dead woman. They had a big load. This boy was scared to death. There was a back room and when the door opened he saw a lot of meat drying in there – dead people's meat. They take this woman's body in there.

There was some mans in there that this boy knows already and a lot of other people that he don't know. They told this boy he is going to be with this people, with the rest of them. They not going to turn him out. There's one man in there like the boss of all. He told the boy he might be a helper of the others. He might give him a job to go out and look for some dead bodies. He might teach him that. He showed him everything that they had inside this room: a lot of meat, a lot of turquoise, a lot of

---

6. A curing ceremonial.

beads – so many you can hardly count them. These beads come from the dead bodies.

[In answer to a question.] They say these people use the meat for food. They think whenever they go out for another person, they eat this meat first so they'll have good luck going.

They told this boy they going to teach him how to use this wolf's skin. They told him there's a wolf skin in there hanging up with the ear cut off. They going to let him use that. They want him to go over the other side of Mount Taylor. They told him to undress his clothes so he did that. He didn't know what they going to do with him next so he just pull his clothes off. They close up other door over here. They told this boy they going to take him inside in this other room and told him he was to have intercourse with this dead woman before he did anything else. And this boy he got scared and the top door was open yet. He asked to go out a minute and make his water. These men said no. But he kept saying this and finally they let him out. He act as if he going to make his water so they go behind him. He ran off. He heard a lot of noise behind him, people behind him trying to catch him. He hid in dead pine tree and he can hear a lot of them running quite close. After these people pass by him he start off the other way. He go a long ways and then he turn to his home. At first when the boy was brought in the witch people didn't know what they were going to do with boy. He was too young, they says. They told him after he'd had intercourse with the dead woman, they'd tell him the rest of it.

Tall Smith from Thoreau was in that bunch of witches. He got killed in a car accident just a few years ago.'

The Hopi Indians are village-dwelling desert farmers in Arizona, neighbours of the Navaho (Ortiz 1979: 514–86). They believe that witches ('two-hearts') are drawn from every race and nation, have a common language, and are organized into a world-wide society. Hopi witches are the leaders of this organization, that holds its conventions underground at Red Cliff Mesa, northwest of Oraibi, one of the Hopi villages. Once initiated into the society witches have to keep sacrificing the lives of their relatives in order to prolong their own lives (Simmons 1963: 120).

Don Talayesva, a Hopi from Oraibi born in 1890, described how, in the 1930s, in a dream, he visited the witches' secret meeting-place (Simmons 1963: 331–3).[7]

---

7. This account and Case VI-1 are from the autobiography of a Hopi Indian. The work was supervised by an anthropologist, but the selection and interpretation of events were mainly the subject's (Simmons 1963: 1–9), although inevitably influenced by the anthropologist's own ideas and the kind of language that he used in preparing the finished book.

Case III-4

In Don's dream his guardian spirit appeared and told him they were going to the secret meeting-place of the Underworld people. He might recognize some old friends, but he should not be afraid because the guardian spirit would protect him. He gave Don a root and told him to chew it, spit it into his hands, and rub it over his body. The root was medicine, and it disguised his appearance.

They flew on a water shield to the top of Red Cliff Mesa, to a cave that led to a *kiva* (an underground ceremonial chamber). Here had gathered the two-hearts of all tribes and nations to enjoy their dances. A pretty young Ute woman wanted to make love with Don, but he resolved to watch the dances. Eight pairs of *kachinas* (masked dancers representing spirits), with strange designs on their masks and carrying yucca whips, danced and sang beautifully. Don and his guardian spirit went outside and put on *kachina* dress and returned to the *kiva*. The two-hearts were surprised to see two strange new dancers and they sang a tune for Don and his guardian spirit to dance to. The songs were the finest he had ever heard. As they danced his guardian spirit warned that the two-hearts planned to make them prisoner, so as soon as the dance ended they rushed out of the *kiva* and floated away on the shield. When the two-hearts pursued, the guardian spirit produced a root which they chewed and spat at the witches, causing them to stumble and fall to the ground. The medicine was Rainbow Medicine, used by witches to keep back the clouds and cause drought. A great rainbow blocked further pursuit, and Don awoke feeling sure the two-hearts would never catch him as long as he was under the protection of his guardian spirit.

The witches' society is a kind of anti-society, based on behaviour and values antithetical to those of normal human society (Mair 1969: 42). Because it attacks the basic moral values of society, the witches' world has been described as an 'inverted' world, the antithesis of human society (Crawford 1967: 120; Lewis 1986: 56; Winter 1963: 292). However, Turner argues that while some of its aspects invert normal behaviour, other aspects do not so much invert human behaviour as caricature it (Turner 1964: 325). We see this in Don Talayesva's account of the witches' dances, that describes a gathering patterned loosely upon normal Hopi ceremonial activities in which masked dancers impersonate spirits. Although witches strike against the moral order their world is an organized one and they have obligations towards one another, just as 'normal' people have to each other in the orthodox world.

The Idea of the Witch

The witch-image of a culture often tells us how and why individuals become witches. Among the Navaho, persons choose to become witches in order to achieve selfish material ends, to obtain vengeance or to become rich, or to give unconstrained vent to their envy by wanton killing (Kluckhohn 1967: 26). In the less condemned form of Navaho witchcraft that Kluckhohn calls 'frenzy witchcraft' practitioners use their powers for success in love, trading, gambling and hunting (Kluckhohn 1967: 40). Among the Shona, although witchcraft is often inherited from a close relative, the recipient has to acquiesce to its receipt. During the earliest stages of the transmission, when the deceased witch appears to her chosen heir in a dream, she can go to a native doctor to be rid of the witchcraft. The doctor transfers the witch spirit into a black goat that is driven away into the bush.[8] However, we are told that novice witches rarely desire to be cured (Crawford 1967: 107–10; Gelfand 1967: 30). Some persons acquire witchcraft by being taught by a living witch (Crawford 1967; Gelfand 1967: 25; see above, Case III-2). Shona witches enjoy their condition. It appears they engage in witchcraft willingly, in order to express their antisocial feelings. The Tangu *ranguma* is a man who expresses the arrogance, greed, selfishness and lust that is in all men, but which normal men control. A *ranguma* may be a man who has deliberately chosen wickedness. Or he may simply be unable to control the destructiveness which is in him. A motive for becoming a *ranguma*, or for confessing to be one, is the desire to be feared (Burridge 1965: 236).

In cultures where people choose to become witches, it is in order to give vent to their selfish and unsocial natures. They wish to pursue selfish, material and evil ends. (In many cultures there exist two categories of occult practitioners: the sorcerer, who kills for gain, but who is less fearsome because his motives are understandable, and the witch, who harms irrationally, for sheer pleasure.) But among many peoples – for example, the Azande, Nyakyusa, and Kaluli – witchcraft is viewed as an inherent and inherited quality, a power over the possession of which the owner has no control and of the existence of which he may be unaware. Indeed, following the influential work of Evans-Pritchard on the Azande, malevolent use of inherent power was taken as *the* diagnostic characteristic of witchcraft in many early works on witchcraft by British anthropologists (for example, Middleton and Winter 1963: 2–4).

But even in these cases an association is made between selfish personal desires and the use of mystical power to promote others' misfortune. The power can only be activated by unsocial desires on the part of the possessor. (Often, the more this inherited power is used, the more

---

8. One of Crawford's informants states that witchcraft is transferred to a male sheep and a black hen (Crawford 1967: 109).

powerful it becomes.) The Kaluli would appear an exception to this general principle. Not only are witches unaware that they have a 'witch-child' in their heart (Schieffelin 1977: 110), but from the account of Kaluli witchcraft it appears that the witch-child behaves independently of its possessor. However, some of Schieffelin's statements imply a linkage between the attitude of an individual and the actions of his witch-child. We are told that the reasons for a witch attack are always trivial, such as retaliation for a minor theft or insult or for an imagined slight. Presumably these precipitating actions are directed against the owner of the witch-child (Schieffelin 1977: 127–8).

Hopi fear that people can be tricked into becoming two-hearts or forced into becoming witches while still children (Simmons 1963: 43, 344–5, 362, 408). The belief that witches can trick innocents into witchcraft, by getting them to attend their meetings for example, or partake of their food, is a common one. However, it seems that once unfortunates become witches, the motives for their behaviour are the same as those of their fellow witches – personal gain and desire to commit evil.

Witchcraft, whether inherited or acquired, is often associated with family membership. Among the Navaho witchcraft is most often learned from a grandparent, parent, or spouse (Kluckhohn 1967: 26). Ceremonial singers and rich persons are suspects for witchcraft, and suspicion is more likely if they are the sons or grandsons of suspected witches (Kluckhohn 1967: 59–60). Beatrice Whiting, studying the Northern Paiute in the 1930s after they had been located on reservations for several decades and had become involved in the national economy in marginal positions, found that both good and bad power appeared to follow family lines. Spouses of known witches were also believed to be witches, and their children came to be so regarded (Whiting 1950: 31). Among the Shona, 'night witches', the most feared kind of witches, usually received their power from a deceased parent or grandparent, and after a witch died she might choose one or all of her children to receive her power (Gelfand 1967: 13, 17–18, 25–6, 30). In Early Modern England, the child or family associate of a known witch tended to acquire her reputation. The Chattoxes and Demdikes were stigmatized as families of witches, and Jennet Device was tried for witchcraft twenty-one years after she gave evidence against her mother and her brother. French peasants of the same period believed that personal power, for good or ill, tended to be passed on within families (Briggs 1989: 26).

A variety of genealogical principles of transmission are recorded for inherited witchcraft. Different cultures may use different principles. Mair, in her 1969 examination of African witchcraft, gives an excellent summary of the variety of lines of transmission recorded from cultures with the belief in hereditary witchcraft (Mair 1969: 47–51). Among the

Azande witchcraft is inherited from the parent of the same sex as the recipient. Among the Nyakyusa a child inherits witchcraft from either parent, but does not necessarily become a witch unless both of its parents were witches. Among some peoples witchcraft is inherited in the line of descent associated with transmission of the main forms of property and with succession to hereditary office. In others it is acquired from the parent other than the one through whom property and office are inherited. So far, studies have failed to demonstrate convincingly any universal social determinants of the line of transmission of hereditary witchcraft (Gluckman 1965: 221–7).

Although witchcraft beliefs are commonly found in human cultures and although the image of the witch often is vicious and vile, members of societies where witches frequently are blamed for causing misfortunes normally are not obsessed with fear of witchcraft. These beliefs only have relevance when some misfortune takes place. Fears are activated and beliefs applied only in certain situations (Crawford 1967: 292–3; Mair 1969: 19).[9] Only when unusually distressing circumstances arise, as with a sequence of misfortunes or an unusually widespread misfortune, are people likely to become obsessed with fear of witchcraft or with the desire to search out witches. (Just as in contemporary Eastern Europe the problems caused by the collapse of the Soviet Union appear to be causing significant numbers of people to become 'obsessively' concerned with seeking out causes or scapegoats.)[10] However, it is argued that the association of witchcraft with such important and emotive events or conditions as birth, health and well-being, and death, often gives witch beliefs a particular emotional intensity (Crawford 1967: 293; Turner 1964: 315).[11]

Symbolic Anthropologists and the Witch-Image

In Chapter 2 I discussed examinations of witchcraft that view it as a product of, or as strongly influenced by, particular kinds of social relationships. Some anthropologists do not analyse witchcraft in this way,

---

9. Briggs says the same for French peasants in the sixteenth and seventeenth centuries (Briggs 1989: 28).

10. For example, in the October 1993 siege of the White House in Moscow, apparently the main topic of conversation among the 'hardliners' come to defend the Russian Parliament was whether or not Boris Yeltsin is a Jew! (*Europhile*, BBC, Radio 4, 9 October 1993.)

11. The personal and social importance of these events and conditions is heightened by the fact that often they are either impossible to control, because some are natural processes, or are difficult to control, owing to limited knowledge or lack of technical ability. Consequently they arouse strong individual and social concern.

but examine it primarily as a set of ideas that create a picture of a particular kind of person. Their concern is with how these beliefs are constituted and they examine the principles on which they are organized. (For example, that the beliefs constituting the idea of the witch are structured upon a principle of opposition to the moral and the normal.)[12] Witch-beliefs are analysed as part of the pattern of ideas that constitutes a culture, and any links between the beliefs and social activities and relationships are treated as secondary, and may be regarded as tenuous. Cultures are often regarded as systems in their own right, and not to be reduced to expressions of society or economics. They consist of interrelated symbols and meanings, and symbolic anthropologists are concerned with how meanings are constructed and with the kinds of messages symbols convey to the members of a culture (Dolgin, Kemnitzer and Schneider 1977; Geertz 1975; Thomas 1975).

These anthropologists correctly emphasize that what superficially may appear to be the same kind of cultural item may have different meanings in different cultures. This can be deduced only by a detailed examination of the item within its cultural framework, where it is part of a wider set of assumptions about the world (Beidelman 1980; Crick 1976; Geertz 1975; Thomas 1975). Some argue that the category 'witch' is a false category. It is created by taking the meanings and beliefs from one culture and arbitrarily applying them to institutions in another (Crick 1973: 18–19; Crick 1976: 117). Alternatively, it may be argued that the definition the analyst applies to a cultural item reflects the theory he holds and that this creates a false field of study (Needham 1978: 28–30).[13] Some anthropologists argue that any kind of cross-cultural comparison is suspect (Beidelman 1980: 34–40; Crick 1976: 159). Consequently they aim to produce analyses of other cultures that translate their concepts and examine the operation of their symbol systems by using intuitive perceptions like those of the literary artist, which require empathy with the actors (Beidelman 1980).

For example, to Malcolm Crick culture is a set of interrelated concepts and norms (standards of behaviour) that define how people ought to behave (Crick 1976: 113). In order to know the meaning of beliefs that anthropologists might classify as 'witchcraft', the relationship between two of the dimensions along which a culture is constructed must be examined. These are its system of ideas about actions and the evaluation of actions,[14] and its system of statuses, with their attributes and powers.[15]

12. For an example from a related occult field, see Stewart 1991: 162–91.
13. For a criticism of this claim see Thomas 1975.
14. Crick terms this the 'system of action concepts and action-evaluation concepts' (Crick 1973: 19).
15. Crick terms these the 'system of "person" categories' (Crick 1973: 19).

Crick believes such examinations will demolish any cross-cultural definition of 'witchcraft'. Beliefs and activities that have been interpreted as parts of a widespread 'witchcraft' syndrome have different meanings in different cultures (Crick 1973; Crick 1976: 109–27). Azande witchcraft and witchcraft in seventeenth-century England are not the same phenomenon, because the two cultures possess fundamentally different philosophical and theological ideas (Crick 1973: 18–19; Crick 1976: 117). However, Crick here seems to be comparing the Azande conception of 'witchcraft' with that of the educated élites of Early Modern England. As I hope to demonstrate, this is a false comparison. Azande beliefs should be compared with the folk belief of Early Modern England, and here there is greater similarity.

Another anthropologist who argues for discarding the cross-cultural category of 'witch' is T. O. Beidelman (Beidelman 1980). In contrast to Crick, however, Beidelman argues for the examination not only of articulated sets of beliefs and ideas, but for a model of examination based upon the relationship between beliefs and activities. In particular, in order to discover the main concerns of a society, the key concepts of its culture and the way they are operated should both be examined. In this way, particular modes of social thought and organization can be discovered and described. To begin the process old analytical categories such as 'witch' must be discarded, so that the categories for understanding the society can come out of the examination itself (Beidelman 1980). (However, many of the features of the persons and practices that Beidelman describes as 'witches' and 'witchcraft' among the Kaguru of Tanzania are those described throughout this book (Beidelman 1986: 138–59).)

Anthropologists who reject the category of 'witch' obviously reject the idea of a cross-cultural image of the witch. What superficially appear to be similarities are dissolved upon a deeper examination of their meanings (Crick 1976: 118). Rodney Needham however, whose concern is with the structure of meanings, accepts the existence of a common image, but denies any common context of application. In different cultures it is evoked in different situations and applied to different sets of relationships. Needham believes our focus should be on the image, as there is no universal context of its use.

Needham argues that culture is largely a consequence of the imagination, rather than of cognitive processes. It is the result of a natural imaginative impulse to symbolize social statuses and their attributes in particular ways (Needham 1978: 47–56). Witchcraft is a complex of images and organizing ideas that occurs almost world-wide. The witch's behaviour is an inversion of normal behaviour and the structure of witch-beliefs is a relation of conceptual opposition – the complex is used to

A Deed without a Name

symbolize the antithesis of morality. Witches are associated with the absence of light, usually with the night. They are represented by colours denoting the abnormal, often black. (As the emphasis is on the *abnormal*, Kaguru witches, who are naturally 'black' skinned, are believed to cover themselves with white ashes, whereas European witches smeared themselves with soot (Needham 1978: 37).) The characteristics of the witch's familiar conform to this opposition to the normal and the moral. Witches are unaffected by normal physical constraints. For example, they may be able to fly.

To Needham, the witch-image is an archetype, a primordial mental image, a psychic constant produced by the human brain (Needham 1978: 45). He argues that such imaginative complexes ('synthetic images') are associated with particular 'social concerns'. In the case of witchcraft the concern is articulated as fear of persons who work evil by secret and invisible means (Needham 1978: 44). The meaning of the image 'corresponds to the concern' (Needham 1978: 42). However, he argues that we cannot say that the concern generated the image. There is no logical reason why this particular concern should produce that particular combination of factors. It is more likely that synthetic images are independent products of the human imagination, of the human brain.

I believe that we can make some generalizations about the context of witchcraft, other than simply the cultural one that it involves a belief in a personally motivated use of mystical harm. However, there are some other problems with Needham's analysis. If it is a primordial mental image, why do some cultures lack a witch-image? It appears to be absent among significantly large sectors of the population in many complex societies such as large industrial states, for example. Furthermore, it may be that the image can be acquired by persons who previously had lacked it, and that this is influenced by social factors. (I would suggest that the Nazi conception of the Jew that took hold in Germany between the two World Wars was an example of this.)

Whilst acknowledging the critical importance of cultural meanings to an understanding of the phenomenon of the witch, and the problems of translation of concepts and evaluations from one culture to another, I would argue that the material that we have at our disposal, flawed as it is, nonetheless allows us to make cross-cultural generalizations about beliefs in the causes of misfortune, and associated patterns of behaviour. The classification of patterns of beliefs and behaviour presents us with problems, and we will never get a complete correspondence between cultures. In most cases we are comparing approximations to a general type, but these may contain enough similarity to our ideal model to enable us to make significant generalizations, that help our understanding of social phenomena. In this respect, our rather imprecise definition of the witch may be a help. It approximates to concepts that appear to be found in many cultures, ideas that have a significant similarity to our concept of evil, involving concepts approximating our concepts of randomness, irrationality, and incomprehensibility of motive. It would appear that this is a real pattern of beliefs and behaviour, found in a large number of cultures.

I do not deny the validity of analyses that focus solely upon relationships between meanings, but I do not accept that they are the only way, and to me they are not the most important way, of examining societies and cultures. In my analysis I am particularly concerned with the influence of power and wealth upon this set of beliefs, with the manner in which they change behaviour towards suspects and influence the interpretation of their supposed activities. I would argue that it is valid to examine belief within its social context, in relation to behaviour. Clearly, action can influence belief. Beliefs about 'witchcraft' may be modified by changes in their social context. In Chapter 8 I will discuss examples of undesired social change in which prophets blame the associated misfortunes upon witches, and where those they accuse are treated much more severely than had been the case in the recent past. The role of power in a situation brings about changes in belief and behaviour regarding the 'witch'. Among the Azande, 'witchcraft' directed against a king is treated differently from 'witchcraft' directed against a commoner. Power modifies the witch-belief.

There appear to be a variety of elements in the idea of the witch, and they may appear in different combinations in different cultures. An institution often shows a degree of variability in its elements cross-culturally, but there may remain sufficient similarity between complexes of associated beliefs and actions for us to regard them as institutions of the same type.

—4—

Detecting the Witch

Amongst peoples who believe in witches there must exist techniques for detecting witchcraft, and usually there are techniques for identifying witches.[1] These are of many kinds, from purely mechanical devices to individuals who have the ability to detect and counter witchcraft.[2]

'Oracles' are one common form of witch detection. In anthropological literature on witchcraft the term 'oracle' has come to mean a mechanical contrivance that, although it may be operated by a human, gives an automatic answer requiring a minimum of human interpretation (Mair 1969: 53–8). The most famous example is the poison oracle of the Azande, as described by Evans-Pritchard for the late 1920s.

The Azande occupy part of the region of the Nile–Congo watershed in Central Africa, their territory straddling the borders of the Sudan, Zaire, and the Central African Republic. Traditionally they were constituted into several kingdoms ruled by a single royal dynasty, the Avongara, who formed an aristocracy. A kingdom contained a number of provinces, each with a governor – a brother or son of the king, or a wealthy commoner. After 1905 the British governed the Azande of the Sudan through their traditional leaders, but the courts of these authorities had lost much of their power and patronage and were no longer the final tribunals of justice.

At the time of Evans-Pritchard's study, some twenty years after the imposition of colonial rule, the Azande of the Sudan were still sheltered from outside influences and pursued an economy based upon cultivation supplemented with hunting, fishing, collecting wild produce, and handicrafts. The colonial government did not recognize the existence of witchcraft and the system of oracles that sought to control it, the linchpin

---

1. The exceptions to this are some societies where witches are outside the community, where there is protection against witchcraft, and witchcraft is divined as a cause of misfortune, but the witch responsible is not identified. Dr Simon Harrison informs me (personal communication) that this is the case in the village of Avatip, Sepik river, Papua-New Guinea. Here, the attempt is made to identify the person within the village who employed the witch, but not the witch himself. This appears to be the situation on the middle Sepik river generally. (See, for example, Forge 1970.)

2. Mair (1969) has three excellent chapters (3,4,7) which make a pan-African examination of this topic.

of which was a prince's poison oracle. When Evans-Pritchard was carrying out his fieldwork witchcraft continued to be an abiding concern for the Azande and they continually consulted local oracles about misfortunes. But they went less frequently to a prince's court to seek the names of witches responsible for deaths, as the verdict of his oracle now had no legal authority (Gillies 1976: vii–xv).

Azande use a number of mechanical contrivances to discover the most propitious course of action or to find the cause of a misfortune. Of these, they believe that the poison oracle is the most reliable. It consists of a poison (*benge*) made from a forest creeper. It has some of the qualities of strychnine, and operates by being administered to a fowl. As the operator administers the poison he asks information on behalf of the consultant. His request is put in the form 'If such-and-such is true, *benge* kill the fowl.' In order to confirm the finding the poison is then administered to a second fowl with the counter-request, 'If the same is true, *benge* spare the fowl.' The poison, prepared in the proper manner and administered in the traditional way, has the power to provide the correct answer to the enquiry. When witchcraft is identified as the cause of a misfortune the victim puts to the poison oracle the names of persons he suspects of wishing him ill. When the oracle identifies one as the responsible witch, he may seek to have the witchcraft removed. The person identified is sent the wing of the fowl killed by the oracle in his name, and is expected to remove his witchcraft by blowing water over it and saying, 'If I possess witchcraft in my belly I am unaware of it; may it be cool (i.e. inoperative). It is thus that I blow out water' (Evans-Pritchard 1976: 59).

Zande believe people die either because they are bewitched, or because they are witches and vengeance has been taken against them for their witchcraft. When someone dies his kinsmen consult a local oracle to discover the cause of death. If it is witchcraft, the names of suspects are put to the oracle. In pre-colonial times the identities of witches responsible for deaths had to be confirmed by a prince's oracle, and the prince could order them to pay compensation or allow their victims' kinsmen to kill them. (The latter was likely only for witches held to have killed several times.) The colonial system outlawed revenge killing, and a victim's kin now must use the traditional alternative, vengeance magic. Its operation is secret and it is believed to act in a wholly just manner. When the deceased's kin believe their vengeance magic has killed the witch they are expected to seek confirmation from a prince's poison oracle.[3]

Other techniques of witchcraft detection require more direct

---

3. For accounts of the Zande poison oracle see Evans-Pritchard 1976; Gillies 1976: xiii–iv; Mair 1969: Chapters 3, 4, 7.

involvement of human intermediaries. Through training and through mystical power, or possession by spirits, they acquire the ability to discover and combat causes of misfortune, including witchcraft (Mair 1969: Chapters 3, 4, 7). Such persons are often termed 'doctors' in the literature as, among other activities, they are concerned with diagnosing and curing sickness. For example, the Shona *nganga* (Crawford 1967: 183–208; Gelfand 1967: 65–158) usually acquires his or her mystical ability through being made ill by a deceased *nganga* – usually a relative such as a father, grandfather, or grandmother – who wishes the sufferer to follow his or her calling. The patient performs a small ritual to inform the deceased that he or she accepts the vocation. He (or she) usually does not undergo any training, but acquires medical knowledge through dreams – for example, about herbs and medicines and how to divine by throwing divining dice ('bones') (Crawford 1967: 190–3). (Some do not divine in this way, but through possession by a spirit.) Or they may be taught their profession by a living *nganga*, often one who successfully treated them for an illness.

A *nganga* may specialize in resolving particular problems. Some specialize in specific ailments such as impotence, barrenness, pneumonia, fits, abdominal swellings, or menstrual problems. Others resolve social problems, such as family quarrels. Treatment often is by sucking out disease, or by making incisions on the patient's body. Medicines may be prescribed for curing illness or repairing social relationships. There are medicines to catch witches or prevent them entering a house. (However, Shona also believe some *nganga* sell medicines to help witches and thieves.)

The most common cases where *nganga* detect witchcraft involve death or sickness, particularly sickness that is severe, swift or protracted, debilitating and painful. Patients or their relatives ideally should consult a *nganga* regarding the cause of illness, but where witchcraft is suspected there is increasingly a tendency to accuse the suspect directly. When *nganga* divine with dice they question both the dice and their clients as part of the process (Gelfand 1967: 105–6).[4] If they divine witchcraft they ask their clients about recent quarrels and about those involved. If they are spirit diviners they identify witchcraft without interpreting any mechanical contrivance, but then ask the same kind of questions of their clients. Doctors often have considerable knowledge of events in their area. (For examples, see Crawford 1967: 189–90.) *Nganga* are often aware that witchcraft is suspected, and discover who the most likely culprits are

---

4. Crawford states that Shona diviners question their clients *before* throwing the dice, but that there is little evidence of audience participation during the actual throwing (Crawford 1967: 191–3).

Detecting the Witch

thought to be. (This does not necessarily mean that they disbelieve their diagnoses.) They then name the guilty witch. The person accused is likely to consult a different *nganga* in the hope of a different diagnosis, and in the past might demand to take the poison ordeal which, it was believed, would cause vomiting in the innocent but would give diarrhoea to the guilty. (For examples, see Crawford 1967: 215–17.)

Witchcraft is not the only possible cause of sickness or death, and different *nganga* may give different diagnoses, as the following two Shona examples demonstrate.

Case IV-1

Gwimbi was 40 when he became ill with abdominal pains in 1958. First he went to hospital, but when his condition did not improve he went to a *nganga*. The *nganga* named a man as responsible, and said he had poisoned Gwimbi's beer because he was jealous that Gwimbi was favoured by their European employer. He said the case was too difficult for him to cure, so Gwimbi attended another hospital where he showed some improvement. Soon, however, his chest pains returned, so he consulted a different *nganga*. This man said that Gwimbi's brother was a witch and was responsible for his condition. Gwimbi did not believe the diagnosis, so he consulted a third diviner, who said he was bewitched by his youngest brother's wife, who was jealous because Gwimbi owned more property than her husband. Gwimbi believed this diagnosis, but got confirmation from a fourth *nganga*. (Gwimbi did not tell him of the previous diagnosis.) He did not wish to take action against his brother's wife, and in 1961 he went back to hospital (Gelfand 1967: 94).

Case IV-2

In 1961 Mia was about 50 years old when he became ill with abdominal pain and diarrhoea. He consulted a *nganga*, who said he was poisoned by someone jealous of his cattle and his four wives. The *nganga* did not tell him who the culprit was, and gave him no treatment. He consulted a second *nganga*, who said his illness was due to the fact that he had had intercourse with one of his wives during her menstrual period. He prescribed treatment, but Mia failed to respond and went to consult the hospital doctors (Gelfand 1967: 94–5).

From Kluckhohn's work (Kluckhohn 1967) it appears that Navaho witches often are identified, and accusations are made, on the basis of individual and community suspicions aroused by the behaviour of the accused, without resorting to a specialist in witch-detection. However, in many cases of illness and misfortune, where the cause of misfortune or the culprit's identity does not appear obvious, recourse is had to diviners. In contrast to other Navaho mystical specialists, diviners do not undergo a long period of apprenticeship to a practitioner. Although they may receive some instruction regarding ritual, their power comes suddenly as a supernatural gift (Kluckhohn and Leighton 1946: 146–9).

Divination is used to locate lost or stolen property, to find water, to discover the whereabouts of persons, to reveal whether one's wife has committed adultery, and (in the past) to predict the outcome of a hunting party or a war raid. But its most extensive use is to discover the cause of personal misfortune and to indicate the proper ritual needed to remedy it. Diviners may operate by 'listening', by star-gazing, or by chewing a narcotic plant, but today the most common method is by 'hand trembling'. The diviner anoints his hand with pollen and says prayers. This causes his hand to tremble, and the way it trembles reveals the information that is being sought. The trembling hand may lead the diviner and his clients to hidden goods, lost property, water, a thief, or a witch. Or the diviner's prayers to the Gila Monster may cause it to tell him what he wishes to know. When a diviner has divined the cause of an illness, appropriate ceremonies may be performed by ceremonial specialists known as 'singers'. For the branch of witchcraft that Kluckhohn terms 'wizardry', where the witch mystically shoots a foreign object into the victim, a specialist 'sucker' sucks out the offending object (Kluckhohn 1967: 34–5).

Case IV-3

'Well, he start hand trembling. He start off with one hand and every time when he rubbed like that [palms together] he start the other hand. He changing from hand to hand all the time. As he hand tremble there he mark on the ground with his fingers. People can't understand what that mark mean, and he pointed off that way [west] many times. And pointed off this way [southeast] and also like he's throwing down something with his hand like that. He got through. When he got through, he sit back over there where his place is [south]. He says, well, mens, this might be truth, may not too.

The way I got it, he says, there's no medicine man sing or pray could cure this girl, but he says who's singing down over there [west] last, does anybody know? They didn't answer for a little while. Then one of the men says, yes, there is somebody singing down there. Who? This man Ugly Singer. He's the one that's singing down there for a man. Says seems to me one or two of you could go down where this man is singing and bring him down here and the way I got it want you to ask the question this man. I think this man did something bad about this girl . . . And another thing he says, I point off this way so many times southeast, something like mile and half away from here. He says this man down there [witch] took something out over there and buried it over there, and that's got to be taken out, right away. Specially if you got this man down there by tonight. Want you people to ask him what he has over there buried. And tell him to change it around and cure the girl. But after he says yes that he did it here, when he says he did all this, trying to kill that girl, if you find out that much, then the girl is going to be safe. If the man doesn't want to tell about it, don't want to make it good for the girl, the girl is going to die right away . . .' (Kluckhohn 1967: 200).

Among the Hopi, again, witches are commonly suspected because of their behaviour. Witchcraft as the cause of illness is diagnosed by native doctors (Simmons 1963: *passim*). The Northern Paiute also may impute witchcraft to specific individuals as a result of their actions. Paiute shamans often diagnose witchcraft when they treat illness, and they may indicate who is responsible. When they do so they may be agreeing a diagnosis already made by the community. They allocate a cause in consultation with those parties interested in the case, and by drawing upon all the known facts concerning the patient and the incidents that preceded his or her illness. Diagnosis is a combined judgement by the shaman and the community upon the patient, and often upon other members of the community (Whiting 1950: 53–4), as in the following Northern Paiute example.

Case IV-4

The Porter family had a bad reputation for witchcraft. In 1934, when they were living at Fort Bidewell, one of Mrs Porter's two boys accidentally shot his brother dead. They were playing with Jane Simms' son and the Porters blamed the Simms for letting the boys play with the gun. Mr Porter was enraged. He cut his arm and smeared his blood around the doors and windows of the Simms' house, so that they would have to cross his power in order to enter. The White Superintendent was informed, and he ordered Mr Porter to wash it off. He warned him that if anything happened to the Simms family he would be blamed and punished. Feelings against the Porter family were running high on the Fort Bidewell reservation, and the Superintendent ordered them to transfer to Burns.

In spite of the family's unpopularity, all the community attended the boy's funeral. Shortly afterwards Mrs Simms' second cousin, Polly, died after a very brief illness. The community regarded her death as an obvious case of witchcraft by the Porters, who were seeking revenge for the death of their son. It was believed Polly had been bewitched by accident by Mrs

Porter. At the funeral Mrs Porter threw a pebble loaded with power at Mrs Simms, but it missed and hit Polly instead. Some informants said the whole crowd saw Mrs Porter throw the pebble (Whiting 1950: 48–9). (If she did so, it was a foolishly aggressive act.)

Mr M. was a cousin of Mrs Porter. He was a mild, shy, conservative man. He was lame, and had a domineering wife. They had six children. Mrs M. drank heavily and had affairs with other men. She and her husband quarrelled constantly, and their fights were witnessed by the whole community. Mr M.'s wife and their oldest son often got drunk together and would physically attack him. He died suddenly, in his early thirties, and Mrs M. claimed he was bewitched by Mrs Porter.

Mr M. had refused to allow Mrs Porter to accompany his family to their camp, and she became angry and told him that he would die soon. Mrs Porter had bad power in snakes, and when Mr M. set out he saw a snake. That night he dreamed it was choking him. Next morning he had a fever, and Mrs M. told the doctor [shaman] who was called in to treat his condition that she believed Mrs Porter had bewitched him. The doctor's diagnosis confirmed her suspicion, and he asked Mrs Porter to bathe and call off her power. She did so, and Mr M. recovered. About a month later he again became ill. A shaman, Dr Thomas, was called in, and he worked on Mr M. for five nights, but he died. Dr Thomas diagnosed witchcraft, but did not name the witch responsible. Mr M.'s family claimed it was Mrs Porter, and his eldest daughter publicly accused her and attacked her with a club. The M. family made life so difficult for the Porter family that the White Superintendent was forced to move them to the Warm Springs reservation for their protection.

Not all the community believed Mrs Porter was responsible. A significant, but minority, faction believed Mrs M. wanted to be rid of her husband so that she could run around with men whenever she wanted, and that she was jealous of Mrs Porter's close relationship with him. Once, when Mrs M. was ill, a shaman, Dr Harry, diagnosed that she possessed bad power given her by a black lizard and it was making her ill. The dissenting faction believed she used this power to kill her husband. It was Mrs M. who interpreted her husband's dream to the doctor, and she may have been afraid of being exposed and have tried to put the blame on Mrs Porter.

Mrs M. was widely disliked. People believed that she and her daughters felt superior because they could always get work in town. However, the Porter family's reputation for witchcraft was well established, and strong enough to override the dislike of Mrs M. and her children. The Porters had already been transferred from Fort Bidewell because of their reputation, and Mrs M. had powerful relatives who were prepared to defend her (Whiting 1950: 33–4, 50–1), so in spite of the dissident minority the community supported the M. family's accusation of Mrs Porter.

Kaluli witches may be identified by their victims or by mediums who have the power to see witches (Schieffelin 1977: 98–103). Unless a witch has caused death his name is not revealed. The sick person or the medium identifies him only by clan and longhouse. A relative of the victim then visits the longhouse and gives a public warning that the witch must desist. If the victim dies, the witch may be identified in several ways. The dying person may speak his name. Or the witch may attend the funeral, where his presence causes contractions of the corpse. Or his identity may be revealed by the deceased speaking through a medium. Once a name has been given it should be subjected to divination for confirmation, and divination is carried out on the basis of all available evidence (Schieffelin 1977: 108).

Kaluli mediums acquire the ability to see witches and cure the sick through marriage to a spirit wife. A witch causes illness by dismembering its victim's invisible body and taking parts away. The medium renews the stolen parts and returns the victim's invisible body to wholeness (Schieffelin 1977: 115, 128, 217, 220).

In sixteenth- and seventeenth-century England witchcraft was often detected, and witches were often identified, by local wizards and wise women. Such persons were termed 'cunning folk', because they 'knew more than other people'. (The term included anyone with special knowledge, and not only persons involved in occult activities (Thomas 1973: 227).) Depending upon their specialisms, they were also described as 'cunning men', 'wise women', 'charmers', 'blessers', 'conjurers' and 'white witches' (Thomas 1973: 210). Their diverse activities included supplying love charms and aphrodisiacs, telling fortunes, predicting the weather and finding buried treasure. However, they were consulted mainly on matters of health and of lost or stolen property (Macfarlane 1970a: 121; Thomas 1973: 253, 277–91).

Although these persons were numerous, they have left few records because their activities were illegal and, in theory, both they and their clients were open to prosecution (Thomas 1973: 292). Alan Macfarlane estimates that in sixteenth- and seventeenth-century Essex no village was more that ten miles from a known cunning person, and in towns such as Colchester there would be several cunning folk. (For the area immediately around Toulouse in France in 1615, Robin Briggs records twelve diviners in seven villages – eight women and four men, including three priests (Briggs 1989: 22–3).) As with African diviners, prestige often increased with distance, and people would go beyond the county boundaries, to

A Deed without a Name

Suffolk or London for example, for a consultation with a well-known cunning person (Macfarlane 1970a: 115–21; Mair 1969: 174–5).[5]

Cunning folk rarely were full-time specialists. Often they were artisans, such as millers and shoemakers, but in villages they were usually of very humble background. (Macfarlane makes the interesting observation that, of 23 cunning men from Essex whose occupations were known, 7 were connected with the medical profession, an indication of the importance given to the supernatural by medicine in Early Modern England (Macfarlane 1970a: 127, 184).)[6] Many cunning folk charged little or nothing for their services, but if their diagnoses later appeared to have been correct they often received gifts. Many appear to have practised for the prestige. But some cunning folk made a substantial income from their activities, and these were often an important source of income for a woman practitioner (Macfarlane 1970a: 122, 126–7; Thomas 1973: 294–8).

Cunning folk carried out their diagnoses by divination. One common method was by the oracle of the sieve and shears, also known as 'turning the riddle'. This frequently was used to find a thief, or some other offender. The points of a pair of shears were pushed into the wooden rim of a sieve so that the sieve hung down from the points. Usually, two persons of opposite sex supported the handles of the shears with the middle fingers of their right hands. The names of suspects were given to the oracle, each accompanied by a phrase such as:

> By St Peter and St Paul,
> If (name of person) has stolen (name of victim)'s goods,
> Turn about riddle and shears and all.

When the name of the guilty person was given, the sieve turned or fell to the ground (Kitteredge 1958: 199; Radford and Radford 1961: 308–9). Another common oracle was a reflecting surface such as a mirror or a basin of water. The client was made to look into the mirror and expected to see the face of the guilty person (Macfarlane 1970a: 125). Other techniques included boiling the sufferer's urine, burning thatch from the house of a suspect, using a crystal ball, and consulting a familiar spirit (Thomas 1973: 219–20).

Like diviners in other societies, cunning persons often had some prior knowledge of their clients' circumstances and suspicions, and familiarized

---

5. Briggs records the same for Early Modern France (Briggs 1989: 24).
6. Macfarlane's sample also includes two schoolteachers and a churchwarden. This is despite the opposition of religious denominations to the activities of cunning folk, which they classified with witchcraft.

themselves further through questioning them. They inquired into recent events and into their clients' relationships. If witchcraft was feasible or was suspected, it might be diagnosed as the cause of misfortune. The diviner then was expected to identify the witch and prescribe action to end the witchcraft. As in other societies, the diagnosis was one acceptable to the client. The cunning person often worked to bring out and confirm the client's suspicions, and left the naming of witches to the clients themselves – as is the case with the mirror oracle. Cunning folk were not regarded as ordinary people, and the heightened emotional circumstances under which a diagnosis was made often helped confirm its validity to the client (Macfarlane 1970a: 121–6; Thomas 1973: 652–6).

Macfarlane gives a detailed account of a consultation between an Essex client and a cunning man concerning a theft of linen. The wizard questioned his client to familiarize himself with his case, then sent him away for nine days while he considered the matter. Macfarlane suggests this gave the cunning man time to make enquiries locally. When the client returned he was made to look into a magic mirror to see the face of the thief. He saw a face and asked the cunning man to confirm its identity, but he refused, saying it was definitely that of the culprit (Macfarlane 1970a: 124). Thomas describes a consultation between an ostler and a wise woman at Fareham in 1579. The ostler developed backache after a quarrel with an old woman. The wise woman told him it was witchcraft, and asked him whom he mistrusted. When he answered 'Mother Stile', he was told to scratch her to draw blood. This would destroy her power (see Case II-5). Then he would soon become well. In another case cited by Thomas, a mother had a wizard consulted about the cause of her child's illness. A servant made the enquiry, and was told 'Your mistress knows as well who hath wronged her as I', thus confirming the mother's suspicions of the name of the responsible witch (Thomas 1973: 656).

To cure witchcraft, the cunning person might prescribe charms to be worn by the victims or spells to be said over them. The client might be advised to make the witch bleed, or touch the victim. Victims or their relatives, or the community if the suspects were widely feared, might take action to drive them out, often by arson. These measures could be taken without recourse to a cunning person. Similarly, the witch might be identified without consulting a witch-finder. Victims might dream of their persecutors. Or someone believed to be possessed might call out a name, which would be interpreted as an attribution of guilt. It might be remembered that the victim, or the victim's parent, had quarrelled with someone who might be a witch (Macfarlane, 1970a: 103–13, 125; Thomas 1973: 628–37, 652–3).

This occurred several times in the case of the Pendle witches. John Nutter told his father that he was bewitched by Chattox and her daughter.

A Deed without a Name

When Christopher Nutter fell ill he said he was bewitched, but named no one. As he lay dying, John Device claimed Chattox had killed him because he had not made her a payment of meal. Other persons often deduced the connection between a quarrel and a death, as in the case of the death of Richard Baldwin's little daughter, and the death of Christopher Nutter. Hugh Moore accused Chattox of bewitching his cattle (for which he died), and John Moore's father accused her of turning his ale sour. Personal relationships and community suspicions must often have made a visit to a cunning person appear unnecessary. Or a case might be taken to law, where the fact that identification already had been confirmed by a 'white witch' could be regarded as important evidence (Macfarlane 1970a: 18, 129).

This latter appears paradoxical. For most of the Early Modern period in England the activities of cunning folk were illegal. Magicians and their clients were liable for prosecution by the ecclesiastical courts, and the witchcraft statutes of 1542, 1563, and 1604 prescribed secular penalties for some kinds of magic. In practice the secular courts tended to be lenient with cunning folk or to ignore them, but ecclesiastical courts were zealous in prosecuting them as 'white witches'. Punishment involved public confession and penance, usually in church before fellow parishioners. Prosecution reached its peak at the time of Elizabeth I. The clergy claimed cunning persons obtained their powers from the Devil. Cunning folk claimed to obtain them from good spiritual sources. However, the proportion of practitioners prosecuted was small, and the popular attitude towards cunning folk was supportive. They were held to perform important services for the community, and this was a main reason why relatively few were denounced to the ecclesiastical authorities. (Some were not accused by churchwardens, and other persons whose duty it was to denounce them, because their occult activities made them too powerful within the community (Macfarlane 1970a: 66–74; Thomas 1973: 287, 311).)

Astrologers also diagnosed witchcraft. Although they claimed that the principles on which their conclusions were based were more formal than those by which cunning folk operated, in practice their system also was highly flexible, enabling the astrologer to take cognizance of the client's suspicions. An astrologer diagnosing witchcraft might give a description of the culprit, but rarely gave a name. Once again, clients were encouraged to allocate blame on the basis of their own suspicions (Thomas 1970: 377, 402–4). As in other societies, we find in English witchcraft a pattern of divination that interpreted events in terms of personal relationships and operated to confirm a client's suspicions. Robin Briggs records the same for the activities of cunning persons (*devins*) in French-speaking Europe for the same period (Briggs 1989: 24, 29).

Strange and unsettling types of behaviour, such as fits or hysteria, have often been interpreted as having a supernatural origin. Christians often have regarded them as attacks by the Devil or demons, who cause mental states in which the victim is not responsible for his actions (Robbins 1959: 392–8; Thomas 1973: 569–87). The demons causing the affliction may have taken over the victim's body. Or his behaviour may be due to his being attacked externally by demons acting upon him from outside his body. The former constituted *possession*. The latter, *obsession* (Robbins 1959: 392; Thomas 1973: 569–70). In Europe both were associated with witchcraft, and the victim often named, or was prompted to name, the person responsible for his condition. In both, the victim can directly accuse the witch. Contemporary popular accounts of possession in Early Modern England and New England present it as the immediate consequence of hostile contact with the individual who brought about the condition. In fact, more reliable contemporary evidence indicates that diagnosis of possession usually occurred following a long period of illness during which the patient failed to respond to conventional remedies. Family and friends, and doctors and clergymen, were likely then to suggest to the patient that his or her sickness was due to possession, and even to propose likely suspects. (Certainly the patient would increasingly be exposed to discussion about witchcraft.) As a result of such suggestion, the patient's behaviour changed to conform to a basic pattern of language and gestures considered diagnostic of possession, a pattern that was defined by doctors and ministers during the course of the sixteenth century (Holmes 1993: 60–5: see especially the cases of Elizabeth Belcher, 60–1; Lady Jenning's daughter, 61; and Mary Glover, 61).

Possession and obsession affected particularly the destitute, the lonely, and those suffering some kind of emotional depression or repression. Its victims usually were women and adolescent girls. Demonologists interpreted this as evidence of the inferiority of women. They were more susceptible to Satanic influence than men (Holmes 1993: 58, 60). The most common behaviours associated with these conditions were abnormal contortions of the face and body, a change in the voice, and vomiting strange objects (Robbins 1959: 392). Possession was a dramatic aspect of a number of bizarre French witchcraft cases from the seventeenth century that involved the possession of whole convents of nuns (Michelet 1960: 159–306; Russell 1981: 87–9). The most notorious of these was the case of Father Urbain Grandier and the nuns of the Ursuline convent at Loudun in 1633–4 (Huxley 1971; Michelet 1960: 189–206; Robbins 1959: 312–17). Contemporary critics of a witchcraft explanation for these cases often interpreted them as due to the repressed sexual condition of the nuns (see, for example, Huxley 1971: 120–1).

In Catholic countries possession was treated by the rite of exorcism,

A Deed without a Name

performed by members of Holy Orders such as the Dominicans, Jesuits, and Capuchins in order to expel the offending spirit. In the case of the Ursuline prioress of Loudun, the demon Asmodeus was finally expelled from her body by an enema of one quart of holy water, during a long and complicated exorcism. The exorcists claimed that he was only one of seven demons possessing her (Huxley 1971: 114–15). After the Reformation, in Protestant countries there was no longer a ritual designed specifically to cure these cases. The church advocated prayer alone. This led to an increased appeal to 'white witches' and magicians, and to the development of lay exorcists (Thomas 1973: 583–7).

The most famous case of obsession in witchcraft in the English-speaking world is that of Salem Village, Massachussets Colony, in 1692 (Bednarski 1970; Boyer and Nissenbaum 1974; Hansen 1971; Robbins 1959; Starkey 1963; Upham 1867). Here, seven or eight girls aged between twelve and twenty years manifested grotesque behaviour, with fits and convulsions. They claimed to see shapes, invisible and inaudible to the unafflicted, in the form of the witches who were causing their suffering. This 'spectral evidence' was the main evidence against the accused, and was provided by the testimony of the afflicted girls. One testified, 'One morning, about sun rising, as I was in bed before I rose, I saw Goodwife Bishop . . . stand in the chamber by the window. And she looked on me and grinned on me, and presently struck me on the side of the head, which did very much hurt me. And then I saw her go out under the end window at a little crevice about so big as I could thrust my hand into' (Boyer and Nissenbaum 1974: 16). The girls exhibited marks upon their bodies, such as supposed teeth-marks, that they claimed were caused by these attacks.

Sometimes a spectre took the shape of a recently deceased person, come to denounce his murderer. Ann Putnam, aged twelve, the most active of the accusers, said she saw the spectres of two women with bloody wounds stopped up with sealing wax. They were the ghosts of the Reverend George Burroughs' first two wives. They told her he had murdered them. The second wife said that Burroughs and his current wife murdered her so that they could marry (Boyer and Nissenbaum 1974: 17). Another of the girls claimed Burroughs' spectre brought her images of people to stick pins into (Robbins 1959: 62). Whenever the accused were brought into court the girls went into fits and claimed they were being punched, bitten, and pinched by the spectres of the accused.

This evidence was supported by evidence of supernatural feats by the accused, and of specific cases of causing harm by witchcraft (*maleficia*). There were nine depositions of acts of supernatural strength against the Reverend George Burroughs (Boyer and Nissenbaum 1974: 12). Rebecca Nurse was accused by Sarah Holton of killing her husband, Benjamin

Holton, three years before. She had wrongly accused him of allowing his pigs to get into her field, and she raged against him and threatened to have the pigs shot. Soon afterwards he was struck blind. Although he recovered his sight he lingered with a painful stomach illness for several months, periodically losing his sight, until he went into violent fits and died (Robbins 1959: 435).

As the girls' notoriety spread they were invited into other communities to investigate sickness. There they were less familiar with the members of the community, and they identified witches through touch. They went into fits that were calmed instantly by the touch of a witch. Those opposed to the accusations and trials suggested that their afflictions were due to demonic possession and not witchcraft. They were being used by the Devil to accuse innocent persons.

Anti-Witchcraft Associations

In some societies there are associations that identify witches. They are characteristic particularly of colonial and developing societies, where disruption and conflict produced by colonization and social change often promote the belief that witchcraft has intensified. From the Nigerian emirate of Nupe we have an account of a witch-finding association from pre-colonial times, *ndakó gboyá* (Nadel 1935; 1954: 163–206).

To the Nupe, witchcraft was a power that had to be acquired, usually through purchase. The power was contained in a medicine that rendered the practitioner invisible and enabled her shadow-soul to leave her body. However, it was the practitioners' motives that made them witches. Witches were motivated by ill will; their motives were not considered normal. They used the medicine exclusively for evil purposes. Their main activity was 'eating' the souls of their victims (Nadel 1935: 425; 1954: 166–7). Witches were of either sex, and in theory a male witch could be as evil as a female one. In fact only women were considered to make 'real' witches (Nadel 1935: 428; 1954: 169–72), and the activities of men witches were less objectionable; but in order to kill women witches must obtain the permission and help of a male witch (Nadel 1954: 178).

Women witches were believed to form an association. Their shadow-souls met at night, each member in turn providing a victim, often a close affinal relative, to be consumed by the gathering. The witches of each town had an official head, the *lelú*, whose office was recognized by the town authorities and by the King. She was elected by the women traders, and confirmed by the town chief – or the King if she was the *lelú* of the witches of the capital, Bida – and was a known witch, but was supposed to use her powers for good purposes only, to discover witches and to restrain their antisocial activities. Her ordinary office was concerned with market affairs and the commercial activities of women (Nadel 1954: 167–9).

Witchcraft was associated particularly with cases of mysterious illness. Where public opinion had focused suspicion on a specific individual the King or the town chief, often assisted by the local *lelú*, tried to make her confess, repair her evil, and pay compensation. More serious cases, such as epidemics, had to be dealt with by the anti-witchcraft association, *ndakó gboyá*.

This was a socially recognized association with an official head, the *májī dodo*, who was from the lineage of its founder and was confirmed in his position by the King. It was organized into local branches, each of which was founded by an individual initiated by a *májī dodo*. Consequently each branch was organized around a local family, the descendants of its founder, but it appears that membership was open to others (Nadel 1935: 442; 1954: 192–5). Members had secret knowledge, enabling them to discover witches. To combat witchcraft they used medicine that had the same qualities as the medicine used by witches. It rendered the user invisible and enabled him to separate his shadow-soul from his body.

The association had the ability to identify witches through its control of a spirit[7] that appeared as a huge dancing mask, a cylinder of cloth about fifteen feet high. The spirit appeared in a village that had become plagued by witches and danced all day in the village square, where all the villagers were gathered. (Several maskers might impersonate the spirit, but only one appeared at any time.) After nightfall the dancing mask identified women witches, who were taken into the 'bush' by the mask's human assistants and forced to scratch earth with their hands until blood came from under their fingernails. This was taken as proof of guilt, and the 'witch' could be warned by the village chief, fined, and ordered to pay compensation for the harm she had done. When the ceremony was finished the spirit disappeared back into the 'bush' whence it came.

Cleansing the local community of witchcraft could be done at the request of the village authorities, who would send a messenger to ask the local branch of the *ndakó gboyá* for assistance. Nadel suggests the messenger conveyed to the association public suspicions concerning the identity of witches. However, at least in the later years of the pre-colonial period and in the early years of colonial rule, the *ndakó gboyá* could, with the approval of the King, initiate witch-hunts on its own initiative. These witch-hunts created so much social disruption and such terror among

---

7. Nadel refers to it as a spirit, but says that its nature is 'more indefinite' (Nadel 1954: 191).

women that village chiefs would come together to collect a sum of money. They would send this to the King and beg him to call off the association. (Because of the disruptive consequences of its activities the *ndakó gboyá* was never allowed to operate as a witch-finding association in the capital. The capital was the King's town and he checked the activities of witches, aided by the *lelú*.) By 1921 the activities of the society appeared to have become so concerned with extortion that it was banned in Bida emirate by the colonial administration, although it continued in the Nupe districts south of the river Niger.[8] However, when Nadel was doing fieldwork in Nupe in the 1930s the people regretted its banning because they believed there was no longer a mechanism to control witchcraft. Consequently a supposed outbreak of witchcraft in Bida in 1932 produced social unrest and led to the illegal killing, by stoning, of three suspected witches (Nadel 1954: 163–4, 200).

In his account of Nupe witches Nadel stresses the role played by public opinion in influencing detection by witch-finders. The village chief and the *lelú* accuse persons on whom public opinion has come to rest its suspicions (Nadel 1935: 433), and Nadel suggests the village messenger conveyed the identity of suspected witches to the *ndakó gboyá* (Nadel 1935: 440). Mair stresses the importance played by public opinion in influencing the detection of witches in traditional African communities. Diviners and doctors elicited information in consultations, and often had a knowledge of local circumstances and suspicions (Mair 1969: 71, 88–9, 95–6). The witch-finder knows whom the community suspects of practising witchcraft. In cases where illness and misfortune is a matter concerning individuals and has not become of concern to the whole community, the diviner or doctor usually confers with his client or his client's representative before making his diagnosis. He discovers their suspicions and the tensions in their personal relationships. His diagnosis is influenced by these factors, and to the clients it is a reasonable one – though not necessarily an acceptable one – for their circumstances. (See the two Shona Cases IV-1 and IV-2.) It enables the clients to take some type of action, often with the continuing help of the doctor or of some other specialist, which it is hoped will remedy their condition.[9] The Shona *nganga* question their clients as part of the process of divining with dice. If they divine witchcraft, they ask them about recent quarrels and about those involved. The Northern Paiute doctor often produces a diagnosis

---

8. Nadel states that as a result of its activities there was practically no money left to pay taxes (Nadel 1935: 442).
9. It is argued that modern doctors (MDs) frequently perform their diagnoses in similar fashion (Hart 1985: 11, 17).

A Deed without a Name

in agreement with one already popular in the community. To reach this he questions concerned parties about the patient and about incidents that occurred prior to the illness. Dr Thomas questioned Mrs M. about her husband's illness, diagnosed witchcraft, but named no one. He had no need to. The reputation of the Porter family was such that the majority of the community already had decided they were to blame. All of this does not mean that doctors, diviners, and oracle-manipulators believe their findings to be fraudulent. As believers in witchcraft themselves, their opinions as to where witchcraft is most likely to be located are usually the same as their clients'.

We have seen that the diagnoses of witch-finding practitioners may conflict with one another, and diagnoses delivered by one authority may be unacceptable to parties involved in the consultation or to members of the public. A client may go to a number of different doctors until one gives a diagnosis that he finds acceptable. Where oracles are used, witchcraft is believed to be such a powerful force that it may interfere with the oracle to produce an incorrect verdict. This may be given as the explanation why different oracles give different verdicts in the same case, or why a judgement that had been accepted later proves to be false (Lewis 1976: 79; Mair 1969: 146). Other reasons given may be failure of the oracle's operator to observe necessary taboos, or other procedural shortcomings on his part.

My impression is that the literature usually emphasizes the ability of witchcraft to influence oracles to pronounce innocent a guilty party, but there is also evidence of mystical interference with respected oracles in order to bring a verdict of guilty against the innocent. In his account of the life of the great Azande king Gbudwe, Evans-Pritchard recounts how the oracle of young Gbudwe's father, the King, confirmed Gbudwe's mother to be a witch (Evans-Pritchard 1971: 287). When Gbudwe was a boy living at the court of his father Bazingbi, one of Bazingbi's children died of sickness. The poison oracle convicted Gbudwe's mother, and Bazingbi ordered her execution. The grief-stricken boy refused to accept her guilt and claimed that she was not a witch, but that her death was due to the jealousy of his father's other wives. (Is this an implication that by their witchcraft they had interfered with the working of the King's oracle? We are not told.) He had an autopsy performed on his mother's corpse. Her abdomen was cut open to see if it contained witchcraft-substance. It did, and her witchcraft was confirmed. When the King learned that the verdict of his oracle had been questioned and his orders disobeyed (and by a boy!) he swore to have Gbudwe's abdomen cut open just as had been the abdomen of his mother, but the prince fled the court until his father's anger subsided. Although the oracle was proved correct in this case, the fact that Gbudwe had the autopsy performed suggests that there was room for doubting the verdict even of a King's oracle.[10] Witchcraft and magic constitute a system of beliefs that mutually reinforce each other. A proven failure of diagnosis can be explained in terms of the system of beliefs of which it is an expression, and it might strengthen, rather than bring into question, the beliefs of those operating within these mutually reinforcing ideas.[11]

Ordeals

In some cultures the guilt or innocence of a person accused of witchcraft might be established by ordeal. In Central Africa in pre-colonial times, Cewa chiefs and village headmen regularly administered to whole villages an infusion made from the bark of the *mwabvi* tree. These communal ordeals also were administered when a series of deaths, or the death of an important person, suggested that witches were particularly active. Depending upon the type of tree used, persons guilty of witchcraft would die or defecate. Innocents would vomit. In the past convicted witches were killed, or enslaved and sent as tribute to a territorial or paramount chief. Although the ordeal is now illegal, Marwick reported it still used in secret in the 1950s on those occasions when an accused person challenged all possible suspects to submit with him to the ordeal. Often this included the accuser himself (Marwick 1965: 87–9).

Among the Shona, if several deaths occurred in succession the village headman might summon a *nganga* and cause the village to submit to an ordeal. Everyone over ten years of age drank poison. People vomited if innocent. If guilty, they had diarrhoea (Gelfand 1967: 107). Among the Nyakyusa accused and accuser drank poison together, and a convicted witch who did not confess witchcraft was driven from the village and often from the chiefdom to which it belonged (Mair 1969: 142–6).[12] Accused persons often demanded the ordeal in order to demonstrate their innocence.

Writing of the poison ordeal among the Lele of Kasai, Zaire, Douglas states that its verdict was regarded as infallible, and it was the only way

---

10. This story calls into question the widely repeated assertion that a prince's oracle was regarded as infallible (see, for example, Lewis 1976: 70). However, Dr Rosemary Harris suggests that among the Azande élite there may have existed considerable scepticism about the manipulation of oracles, since they had a wider knowledge of the verdicts given in cases of death (personal communication).
11. See the case cited by Mair, p. 72 below. The way such systems of belief are self-reinforcing is the topic of Luhrmann's examination of occult groups in contemporary England (Luhrmann 1989).
12. Mair deals with other ordeals, as well as the poison ordeal.

people who acquired reputations as witches could clear themselves. They would welcome the ordeal as the means to end their troubles and demand compensation from their accusers. (Whatever the verdict, it would terminate the factionalism that had grown up around an accusation of witchcraft.) When the Belgian administration of the colonial Congo effectively prohibited the ordeal these unfortunates had no way to clear themselves and were driven from the community. New anti-witchcraft movements had to arise that could take over some of the functions previously performed by the poison ordeal (Douglas 1963: 123–41). However, Douglas' claim that the Lele ordeal was believed infallible is challenged by evidence from other cultures. Just as witches may be so powerful that they can interfere with the verdicts of oracles, so they may tamper mystically with the operation of an ordeal. A Cewa informant told Marwick that when a witch knew he had to drink *mwabvi* he might shut a cock in a basket and leave it in his hut. When he drank, the cock would defecate in his stead (Marwick 1965: 88). Where ordeals are used to prove guilt or innocence, people may believe their operation can be perverted by witchcraft. Mair cites a case noted by an anthropologist in northeastern Zambia, apparently in the 1960s. A man was alleged to have put his arm in boiling water as an ordeal to prove his innocence of witchcraft. Although it was done only a few days previously, there was no mark. He claimed this demonstrated his innocence. Others said his witchcraft was so powerful that it had defeated the ordeal! (Mair 1969: 146).

King James I of England[13], who considered himself an expert on witches (Clark 1977), believed the most conclusive proofs of being a witch were provided by 'fleeting' (floating) a suspect on water to see if she would sink, as water refused to accept a witch, and finding the Devil's mark. Neither was resorted to in the case of the Pendle witches (Peel and Southern 1969: 23, 35–6).[14]

---

13. King James I of England was simultaneously James VI of Scotland. (See Case X-1.)
14. For an account of a Continental search for the Devil's mark, that of Urbain Grandier, see Huxley 1971: 155–6. See also Macfarlane 1970a: 19.

—5—

Witch Suspects: Suspicions Based upon the Personal Qualities of Suspects

The next four chapters discuss the kinds of persons who are accused of practising witchcraft. World-wide, we find that the social distribution of witch suspects is never random. They are drawn from a few specific categories of persons, although the membership and importance given to particular categories varies between societies and cultures. (For example, the composition of socially deprived or marginal categories may differ between societies, and marginal persons may form an important category of suspects in one society but be assigned less importance in another.) Consequently the pattern of accusation is not everywhere identical. Membership of categories is especially influenced by the structure of society and the operation of the economy. The importance of particular categories in producing witch-suspects varies with beliefs about the nature of power, and other aspects of ideology, which influence witch-beliefs and appear to have some independence of economy and social structure. (They may determine the sex and relative age of witches and the line of transmission of witchcraft, for example, and consequently they can influence the attribution of harmful powers to particular types of persons.)[1]

Studies of African witchcraft have emphasized accusations reflecting tensions and strains within the community, giving the impression that the majority of accusations in African society involve persons in social relations in which there are conflicting interests. On the other hand, Melanesian studies generally stress accusations and hostilities between communities. Some of this difference may be only apparent, a consequence of fieldwork methods and types of information collected. But there does appear to be a genuine difference between the two areas, with the emphasis in African societies tending to be on internal witchcraft and with Melanesian societies emphasizing external witchcraft (Marwick

---

1. For general discussions of the difficulty of determining a general relationship between witchcraft beliefs and social factors, see Gluckman 1965: 222–7; Lewis 1976: 98–105.

1970). Why this should be has yet to be answered satisfactorily. However, I am informed that in Melanesian societies where the witch is the stranger from without, living outside, he is still employed by someone within the community. There is still an 'enemy within' who is associated with witchcraft, although he may not himself be the witch.[2]

One convenient way of discussing suspects is to divide them into those who are suspected on the basis of personal qualities they are believed to possess, and those who occupy social positions and statuses the demands of which promote tension with persons with whom they are in particular social relationships. These two divisions overlap and may mutually reinforce each other. For example, although the elderly often are accused because the demands associated with their position promote hostility with some of the persons with whom they have relationships, at the same time many behave oddly or eccentrically, which may arouse suspicion about their behaviour.

Why members of particular categories should become suspects will be discussed as we examine each category in turn. In this chapter I wish to deal with persons suspected because of their personal qualities. Of these, a large and important category consists of persons who detect or cure witchcraft or identify witches.

Witchfinders and Curers

Nadel says of Nupe witchcraft, 'In order to . . . fight witchcraft one must possess witchcraft oneself' (Nadel 1935: 436). He is referring to the fact that the medicine used by members of the *ndakó gboyá* has the same properties as that enabling witches to carry out their nefarious activities; and that the *lelú* is a convicted and repentant witch. The idea that repentant witches may turn their abilities to the use of the community appears widespread. The Nyakyusa believed they made good defenders of the community from witchcraft (Mair 1969: 153), and some Tangu *ranguma-*killers were 'retired' *ranguma* who had been wicked in their past but now were using their abilities for the defence of good men (Burridge 1965: 233).

Human witch-detectors often are believed to possess the same kind of power as the witches they identify and combat. How otherwise would they have this ability? This has been reported for many African societies. The Tanzanian Safwa believe people can possess a power, *itonga*, which is neutral in itself but can be used for either good or evil purposes. However used, it is with conscious intent, and the possessor is morally responsible for its actions (Harwood 1970: 57–9). Most Zande believe

2. Dr Simon Harrison, personal communication.

witches and 'witch-doctors' have an identical power, *mangu*, the possession of which enables Zande doctors to discover the activities of witches. Again, it appears to be a neutral power that can be used for good or evil ends. The Lugbara of Uganda believe a single power, *ole*, can be used for social or antisocial ends. A similar situation exists among many East African peoples (Harwood 1970: 69–76).

Since human witchfinders often possess powers identical to those of witches, it is not surprising that they may be suspected of using them for their own ends. Just as witches may become witch-finders, so are witch-finders likely to become witches. Persons who cure witchcraft may also be suspected of witchcraft. They too have special powers and are associated with the world of witches, and as a result of their profession they may possess other characteristics that leave them open to suspicion and accusation.

We can illustrate the ambivalence regarding witch-finders with data from the Safwa, a dialect-group of some 65,000 persons living in Tanzania. They are swidden cultivators, with agriculture carried out mainly by women. In the 1960s, when the fieldwork was carried out, different Safwa tribes were involved in the cash economy to different degrees, as migrant wage labourers on the Copper Belt of Zambia or in the more distant parts of Tanzania. When a man acquired enough money to obtain a second wife he usually returned home and remained there. A Safwa tribe comprises a number of communities linked by patrilineal ties between their headmen. Communities contain a number of compounds whose members jointly work their fields under the direction of the community headman (Harwood 1970: 1–29).

Safwa believe in a mystical power they call *itonga*. It is possessed by some persons, male or female, and transmitted from father to children. The life-force of the possessor carries out mystical activities, invisibly, at night. *Itonga* may be used for good or evil. It is used for good in divination, and to protect the community from attacks by persons using their *itonga* for evil. Like the neighbouring Nyakyusa, the Safwa community has a league of elders who are believed to use their innate power to protect the community. When disputes occur between communities, they use it at night to attack their opponents from the hostile community. *Itonga* may be used antisocially to induce foreign substances into another person's body or gardens. This happens when the possessor of *itonga* feels anger but fails to express it openly. Or it may be used, at night, to consume people, and may kill them. Possessors of *itonga* can tie women's wombs to prevent conception, and tie men's intestines to prevent defecation. They can pervert people's natures. But they can only 'consume' members of their own patrilineage. To kill non-members one must operate in conjunction with some of their patrilineal kin, and this creates blood debts that require the deaths of one's own patrilineal kin in repayment. However *itonga* is used, it is believed to be done consciously. In order to use one's *itonga* for evil purposes one has only to wish to do so. Once it has been used for evil, its possessor must drink medicines or make a public avowal of a change of heart in order to use it for good. Should he then revert to evil, he will die. Attack by *itonga* is repelled by the good use of *itonga* and by medicines prepared by diviners, who can divine the effects of *itonga*. But diviners may also prepare maleficent medicines (Harwood 1970: 57–67).

Case V-1

The headmen of the communities of Igamba and Ibala maintained a close, co-operative relationship. They were related through their common patrilineal grandfather. To outsiders the two communities formed a common unit, Itete, of which the headman of Ibala was the senior elder.

Some residents of Igamba began cultivating unused fields belonging to Ibala, without permission. The matter was brought to the two headmen, who ruled that the fields must be returned after harvest. However, some Igamba residents continued to cultivate Ibala fields. Soon the wife of Mpεnza, assistant headman of Igamba, died. The native autopsy revealed she had died of *endasa*, an affliction caused by being 'speared' in nocturnal battle with a person possessing *itonga*. It was believed to have been caused by the defenders of Ibala in their mystical night fights to protect their fields against Igamba. This ended encroachments on Ibala's land, but deaths continued in both Igamba and Ibala. These also were divined as *endasa*, and it was believed that the defenders of the two communities must be engaged in a night war over the incident of the fields.

In spite of this the two headmen maintained their close relationship and endeavoured to end the dispute. Mwamatete, the son of Mpεnza, was the sixth death attributed to *endasa* in Igamba alone. Cause of death was divined by Chikanga, a well-known diviner, whose verdict was that Mpεnza and his half-brother were in a 'league' of *abitonga* (i.e. possessors of *itonga*) which had demanded Mwamatete's life. Mpεnza and his half-brother were responsible. Anxious to end the dispute, the two headmen interpreted Chikanga's divination to mean that the defenders of Igamba had called in outsiders (including Mpεnza's half-brother) to help them in their mystical night war against Ibala, thereby incurring a debt which they had been asked to repay by the death of a resident of their community (Mwamatete).

Faced with this image of their own self-destruction the people of Igamba formally agreed to stop planting Ibala fields, and the two headmen arranged a ceremony to their common ancestor to confirm this (Harwood 1970: 84–91).

In this interpretation we observe the defenders seeking to protect their community from mystical attack by utilizing the same powers that are used by evil persons. In the process they must indulge in dubious actions, causing them to sacrifice an innocent member of the very community they are protecting.

Among the Navaho, attitudes towards ceremonial 'singers', who perform rituals to cure persons who are ill, are ambivalent, and they are subject to suspicion and malicious gossip. Ceremonial chants have great power, intended for good. But many Navaho believe they also can be used to bewitch, and that an evil singer will use them for evil ends. Any singer is believed to know some witchcraft, in order to defend himself against witches (Kluckhohn 1967: 47). Frequently, successful ceremonial practitioners are accused of witchcraft. As they are powerful and may be rich, jealousy is often a factor promoting suspicion; but Kluckhohn suggests there may also operate in Navaho culture a distrust of extremes – too much of anything arousing suspicion. (Thus those who are 'too poor' are likely to be accused; just as are those who are 'too rich'.) Suckers, who suck witch-objects out of a patient in order to cure that branch of witchcraft Kluckhohn translates as 'wizardry', are regarded with particular scepticism. They generally are believed either to be witches or to be in league with witches, and are feared. Many are believed charlatans. We have already mentioned the Navaho belief that witches engage in fee-splitting. A curer works in partnership with another witch, who causes an illness that is then cured by the practitioner. The two witches split the fee between them. In this way both become rich.

Case V-2

'The Ghost Dance never did any harm, but the witch business did. After Sumner[3] the Navaho were very poor. The government gave them rations for ten years. Some of the families had a few sheep and they got a good start right away. Such families, headmen who were good talkers, young men and healthy girls would all of a sudden get sick. The chanters were not able to help them. So they got suckers to suck these suffering persons. They would always suck out something, a deer hair, a piece of charcoal, a piece of bone, a wildcat's whisker or a porcupine quill.

3. From 1863 to 1868 the Navaho were confined to a reservation at Fort Sumner, New Mexico, by the US army.

As soon as this was sucked out the patient would feel better. They would say to the sucking doctor what is this and the doctor would say this was shot into the body by some witch. Then they would lay this object aside and think no more of it and the doctor would go home. But the patient would not be altogether well and after awhile would have a relapse. Then the family would get another sucking doctor who would suck out something else. This kept on until the people began to say, "I wonder who is doing all this?" Of course the sucking doctors got big pay for this.

They would ask the sucking doctor, "Do you know who shot this?" and he would say, "Yes." Usually he would say this belongs to the first doctor you had. The family would say, "Can you shoot it back?" The doctor would then shoot it back for a big price. Then star-gazers were hired to find out who did the shooting. Then hand tremblers who figure out the signs which their hands are giving them when trembling were also hired to find the guilty ones. When the family thought they had enough evidence they went and got the guilty party. He was forced to give a ceremony to cure the patient. Later it got so bad that they were killing those doctors.

It was not until this time that the sucking doctors sucked out objects. Before they just sucked out blood. These doctors got so numerous that they called themselves witches and offered to bewitch people in different ways. They got saliva, a hair of a person or any belonging that a person had touched and buried it in a grave and made people sick. Of course the people couldn't stand this so they started killing these doctors off. The agent at this time was as ignorant as the people themselves; he said that if you find a witch kill him. Finally, the government stepped in and stopped it. They used to get a witch and they would not let him eat, drink, urinate or sleep until he pointed out someone who was responsible for causing the sickness. It got so bad that no medicine man was safe' (Kluckhohn 1967: 193–4).

Among the Apache, who were neighbours of the Navaho and are part of the same group of Apachean peoples of the southwestern United States (Opler 1983a), both witches and 'medicine men'[4] possess supernatural power. Because women's power is usually weaker than that of men, most witches are men (Basso 1969: 56). They use their power privately, to cause sickness, death, and certain forms of insanity. In contrast, medicine men use their power publicly to conduct ceremonies, and to diagnose and cure illness. Medicine men have unusually strong power, and because they can

---

4. This is the term Basso uses in his analysis of Western Apache witchcraft (Basso 1969). Opler uses the term 'shaman', as they acquire their power through visions (Opler 1941: 242–52 and *passim*; Opler 1983b: 416).

cure witchcraft-disease it is assumed they also can cause it. Through anger, a medicine man may use his power privately for an evil purpose and so become a witch. Consequently, although medicine men are held in high esteem, they are also feared (Basso 1969: 29–40).

Basso, in a study of the contemporary Western Apache of the Fort Apache Reservation in Arizona, collected details of 27 witchcraft accusations that occurred between 1960 and 1967 (Basso 1969). Of these, two are against medicine men. Basso's data are unusually detailed for material on North American Indian witchcraft, and we will use them a number of times.

Case V-3

X was a medicine man who possessed bear power and snake power. In 1962 Y had a bad skin rash and complained of dizziness. He paid X to conduct two ceremonies for him, but his condition did not improve. At X's recommendation a *gan* dance[5] was performed at considerable expense to Y's family and the members of his clan. Still his condition did not improve.

People began to say that X's power was weakening and his fees were too high. In 1964 he left Cibecue community to live with his sister's family in East Fork. He returned in 1965 and performed two bear ceremonials for Z, who had chronic diarrhoea and chest pains. When they failed to effect a cure, rumours circulated that X had learned bad power in East Fork, and may have used it on Z in order to be paid to cure an illness he himself had caused. Z did not accuse X of witchcraft – in fact he took special care not to offend him – but X was not asked to conduct any more ceremonials and returned to East Fork. He told Basso that the people of Cibecue always talked bad about people, so he had gone to live in East Fork, where he would soon carry out a bear dance (Basso 1969: 63–4).

In England, cunning folk who became unpopular with their neighbours might be accused of witchcraft. This was especially true of 'wise women', because they were more likely to treat witchcraft cases, whereas cunning men were more often concerned with lost goods and healing. Consequently wise women more often appeared at secular courts accused of maleficent magic than did cunning men, who appeared more frequently at ecclesiastical courts. The wise woman might be respected for her knowledge and have many clients, but she was always in an ambivalent position because of her occult powers. Thomas cites the cases of Ursula

---

5. A curative ritual in which men represent the mountain spirits (*gáhan*) (Dutton 1975: 70–1).

Kempe, who had been a 'white witch' and was tried at the Assizes in 1582 for 'black witchcraft', and Margary Skelton, prosecuted for evil witchcraft at the Assizes in 1572, who had been tried six years before in the ecclesiastical court for 'white witchcraft' (Thomas 1973: 318, 654).

Of the Pendle witches, Chattox and Demdike, and possibly Elizabeth Device, practised as 'wise women'. Old Demdike was asked by John Nutter to cure his sick cow, and by Moore's wife to cure milk that had gone sour (Peel and Southern 1969: 125–8). The line between maleficent witchcraft and beneficent magic was a very fine one. Many persons (such as Chattox and Demdike) must have been thought to practise both at the same time, or were regarded as different types of practitioners by different persons. Thomas even notes a belief like that among the Navaho, that witch and cunning person may work in partnership (Thomas 1973: 656)! However, Macfarlane states that in Essex there appear to have been few cases of cunning folk accused of witchcraft. Out of over 400 accused 'less than half a dozen' are known to have been cunning folk (Macfarlane 1970a: 128).

Briggs says that a significant number of magical healers (*devins*) were among those accused of witchcraft in Early Modern French-speaking Europe. The power to injure and the power to heal were regarded as virtually inseparable, and the same person might be identified with either in different circumstances (Briggs 1989: 15–16, 24, 26).

One of the most extreme examples of curers being identified as witches must be the situation that pertained among the Mohave Indians. Today the Mohave are divided into two groups, one living in the town of Needles on the California side of the Arizona–California border, and the other sixty miles further down the Colorado river on the Colorado River Reservation in Arizona and California. In aboriginal times they lived along a 150-mile stretch of the lower Colorado river, at the junction of California, Nevada and Arizona (Stewart 1983: 55–70). This is an arid region with mild winters, very hot summers, and very low rainfall; but along the Colorado annual flooding had created lush oases, supporting a relatively dense population. The Mohave and other Yuman-speaking tribes of the Colorado and its tributaries planted maize, beans, pumpkins and melons in the flood silt of the river, and supplemented them by fishing, hunting, and gathering wild plants. Each tribe occupied a string of sprawling settlements along the river valley. There was little in the way of government, but tribal identity was strong, and overrode identity with the local community.

Mohave believed that all special talents, skills and successes were the consequence of dreams. All important persons – chiefs, outstanding braves, shamans, singers, and funeral orators – were believed to have acquired their positions because of power given to them in dreams. These came first to the unborn child. They were forgotten at birth, but dreamed again later in life. Shamans had the most elaborate dreams, enabling them to cure illnesses; but they could cause disease as well as cure it.

In 1859 the Mohave were defeated by the US army, and there followed a period of social and cultural change. From 1870 to 1890 was a particularly demoralizing time; but their condition began to improve around the beginning of the twentieth century.

Although the Mohave were concerned about witchcraft, they were not obsessed by it (Stewart 1973: 315). Witches were either shamans or persons with the power to sing certain song cycles. It was believed that in every generation some individuals, having dreamed certain dreams in their mother's womb, were predestined to become witches. Not all shamans were witches, but shamanistic power could be used for good or evil. Mohave believed that if one could cure a specific disease, one could also cause it. Consequently all doctors were regarded with suspicion, especially when they got older, as it was believed they became more inclined to use their power malevolently.

Most witches were men, but women witches were the more dangerous. An unusual feature of Mohave witchcraft is the belief that a witch cannot harm anyone he dislikes. (Consequently Mohave behaved meanly towards persons they suspected of witchcraft.) A witch's victims were usually persons he liked. Often they were comely young men and women, and were often his relatives. But he might bewitch someone with whom he was angry or of whom he was jealous. He might bewitch a professional rival or someone who disparaged his powers, and occasionally he might be hired to kill someone for payment. Witches liked to kill.

The souls of the witch's victims were sent to a special place where he visited them in dreams to have sexual intercourse, which often was incestuous. Should the witch die a non-violent death he would lose his retinue of souls, who were liberated. If he died violently he retained his control over them in the Afterworld. Or a more powerful witch might kill him and capture his retinue of souls for himself. Consequently he might provoke his own assassination, and Mohave believed that ultimately the witch succumbs to the temptation to be with his beloved victims and provokes his own murder. Every murder of a witch was seen as deliberately provoked, and therefore a form of suicide. The witch who died a natural death cried on his deathbed, because he was losing his retinue of beloved souls.

It is reported that shamans were killed for blighting crops, killing their patients, informing a sick person that he would die, and admitting responsibility for deaths. The Mohave believed killing witches was socially beneficial, and in aboriginal times braves openly were encouraged to kill witches who had become a public menace and were threatening the welfare of the community. Witches were said to be unable to control their behaviour. They would do anything, including committing incest and bewitching their relatives. It was said of brave men that they were like witches (and shamans in general) in that neither wished to live long. Several recorded witch-killers, who took it upon themselves to beat to death suspected witches, were themselves shamans and witches, and ultimately would have been killed had not United States laws been enforced on Mohave society.

It appears that after their 1859 defeat the stresses and problems produced by United States conquest and control led the Mohave to believe that witchcraft had increased. Witch-killing became rarer after the end of the nineteenth century, the last recorded instance being in the 1930s. Contemporary Mohave disagree over whether the legal prohibition on killing witches has caused witchcraft to increase – in order to pursue dream intercourse – or decrease – because now witches die a natural death they can no longer keep their retinue of beloved souls (Devereux 1961: especially 371–425; Stewart 1973).[6]

Case V-4

Sahaykwisa: was born in the middle of the nineteenth century and was killed near its end, at the approximate age of forty-five. She was a lesbian transvestite, and may have undergone the Mohave initiation rite for female transvestites. Her appearance was feminine, but she professed to be a man. Occasionally she prostituted herself to Whites. She was an industrious farmer and hunter and a practising shaman in venereal diseases, which made her lucky in love. She began to practise witchcraft in her middle twenties, and was first accused of witchcraft about five years later.

Sahaykwisa:'s first wife was very pretty, and her male suitors ridiculed her for living with a lesbian. Finally she eloped with a suitor, but returned to her lesbian husband. Sahaykwisa: took her to dances and sat with the men and boasted of her wife in a typically masculine manner. People teased her wife over the relationship until she complained to Sahaykwisa:. Sahaykwisa: became angry and told her to go, and she left.

Sahaykwisa: found another wife, who was also subject to a great deal of teasing and ridicule. Sahaykwisa: and her current wife met the former wife and her new husband at a dance. The former wife ridiculed Sahaykwisa:'s current wife while the men present egged them on to fight, and meanwhile Sahaykwisa: and the husband maintained the dignified bearing assumed by men when women fight over them. The crowd jeered Sahaykwisa:, and a practical joker pushed the fighting women on top of

---

6. For an account of Mohave cultural persistence and change, see Gorman 1981.

her and they all rolled in the dust. Soon the second wife had had enough of the insults and left Sahaykwisa:, who painted herself as men do when going on the warpath or going to fight their wife's seducer. It was assumed she was going to fight her eloping wife's paramour, but instead she visited a woman in another camp and seduced her into sleeping with her. The woman's husband, Haq'ua, did nothing, as 'he could not very well fight with a transvestite'.

Sahaykwisa: earned enough to provide her new wife with quantities of beads and pretty clothes. In spite of this she returned to Haq'ua, who accepted her back with some reservation because she had lowered herself to become the wife of a lesbian, and because Sahaykwisa:, who was now recognized to be a witch, might bewitch him. She had killed several women by witchcraft, and cohabited with their souls in her dreams. Now she went to bewitch her third wife, but people warned Haq'ua and he waylaid Sahaykwisa: and raped her.

After this Sahaykwisa: ceased to court women. She became a drunkard and developed a craving for men, and she was often subjected to gang intercourse when she was drunk – a common Mohave action towards drunk women. She fell in love with an elderly man, Tcuhum, of her own clan, but he refused to cohabit with her 'because she was a man'. Sahaykwisa: bewitched him in order to have intercourse with his soul in dreams. As Tcuhum died without revealing the name of the witch who was killing him, people assumed he wished to become its victim. (This was the only person she is reputed to have killed who was a relative.)

By now Sahaykwisa: longed for the company of those she had bewitched, and was looking for a chance to get herself killed. She started an affair with Tcuhum's son and his friend. Sahaykwisa: got drunk and boasted to her two lovers that she had killed Tcuhum, so they threw her into the Colorado river and she drowned. Her body was found and cremated – the Mohave practice with the dead. When people learned she had been murdered they did nothing, because witches should be killed and wish to be killed (Devereux 1961: 416–25).

Antisocial Persons

Witchcraft is associated with evil feelings. It is associated with jealousy, quarrelling and antisocial behaviour. Consequently persons who consistently behave antisocially may be in danger of becoming identified with witchcraft and accused of being witches. Anthropologists who follow a functionalist approach have used this fact, and related aspects of witchcraft complexes, to emphasize that witchcraft sanctions a code of morality. Witchcraft beliefs dramatically focus attention on moral values, and functionalist anthropologists argue that consequently they are a force for conformity.[7] They point out not only that unsocial persons may be accused and that this is a pressure to conform, but also that fear of offending a possible witch may be another reason for behaving with propriety. If one behaves unsocially towards witches they will be offended, and an offended witch is a dangerous enemy.

Max Marwick analysed 101 witchcraft cases[8] that he collected among the Ceŵa of Central Africa in the 1940s and 1950s, covering a period going back to the beginning of this century (Marwick 1965), and deduced that in about half the accused manifested antisocial behaviour prior to being suspected or accused (Marwick 1965: 124). He also deduced that in about 60 per cent of his cases the victim of witchcraft, or a close associate, had, before the attack, been guilty of a misdemeanour directly related to the attack (Marwick 1965: 282).

The Ceŵa live in eastern Zambia, Malawi, and Mozambique. They numbered about three-quarters of a million at the time of Marwick's research, and were hoe cultivators, growing maize. In the mid-nineteenth century many Ceŵa communities came under the control of the militaristic Nguni, who had fled northwards to escape Shaka's Zulu power in southern Africa. At the end of the nineteenth century Europeans broke Nguni power, and the Ceŵa were made subject to European domination and influence. Increasingly they were forced into a money economy in order to pay taxes and acquire European goods, and the major source of income became male migrant wage labour.

The village is the basic unit of Ceŵa social organization. The average village contains about sixty huts and has a population of just over one hundred persons.[9] Villages usually contain a number of residential sections, which are kinship units with a matrilineal core. Each is headed by its senior member. The village headman is head of the founding matrilineal section. In the indigenous political system he was under the jurisdiction of a territorial chief, who was subject to a paramount chief. The Ceŵa were divided into many such chiefdoms. By the time of Marwick's research in Zambia and Malawi (then the British colonies of Northern Rhodesia and Nyasaland) Indirect Rule had entrenched this

---

7. See for example, Gluckman 1959: 86; Gluckman 1965: 222; Marwick 1952: 124. However, we should note Nadel's comment that, among the Nupe, some antisocial characteristics may come to be attributed to a suspect *after* she has come to be suspected (Nadel 1954: 170–1). This must be a common rationalization in many cases where a person has come to be suspected for other reasons. See for example, Macfarlane on English witches, where being a good neighbour may later be interpreted as evidence of hidden antisocialness (Macfarlane 1970a: 172).

8. By 1965, Marwick had come to term all these cases 'sorcery' on account of the techniques used (Marwick 1965).

9. Ethnographic background to the Ceŵa is taken from Marwick 1952: 130–4, 215–16.

system, but it also had brought about important changes. Chiefs judged disputes between members of different matrilineages, but within matrilineages most types of dispute were settled without recourse to the colonial legal system.

The Cewa believed a 'good person' was sociable, hospitable, mild and self-controlled. He or she was sexually potent and fertile, had many friends and relatives, took care of matrikin, and was respectful to social superiors. A 'good man' bought clothes for his wife, and if he was a polygynist he distributed his attentions fairly between his wives. The 'good woman' was hospitable, mild, self-controlled, and obedient and sexually faithful to her husband.

The Cewa believed people could be witches who harmed or killed their fellows by harmful magic or other supernatural means learned during childhood. They could be of either sex. Cewa believed witches were usually women, but in fact they made accusations against men and women in equal numbers. They distinguished two types of witch. *Mpheranjiru* were motivated by hatred. *Mfiti yeniyeni* were driven by a hunger for flesh. Witches usually attacked their matrikin. They operated at night, had familiars, and formed a necrophagous guild. In precolonial times suspected witches were subjected to a poison ordeal. When colonial governments suppressed the ordeal, divining became more important as a witch-finding technique.

Case V-5

Gombe was a Cewa village headman. When his sister's child died, Jolobe, the headman of a section within Gombe's village, did not attend the funeral. Instead, he went off to a neighbouring village to drink beer. Nonetheless, when he returned he asked for a share of the funeral porridge. Because of this antisocial behaviour Gombe and his friend Blahimu, the headman of another village, suspected Jolobe of killing the child by witchcraft, and they 'closed' its grave with a magic snare. When they came to examine the grave they found hyena tracks which came up to the grave and vanished at the snare. Jolobe became mortally ill, and before he died he abused his friend Ngombe for misleading him into believing the grave was not closed and they could go and eat the corpse. Jolobe asked Gombe to neutralize his magic, but he refused. After Jolobe's death one of Blahimu's relatives died. Gombe and Blahimu again set a snare, and again hyena tracks vanished when they reached the grave. Ngombe became ill and asked Gombe and Blahimu to remove the effects of their magic, but they refused and he died.[10]

10. This, of course, is Gombe's account.

At the time Jolobe died Gombe's elder sister Tapita recovered from an illness. Members of Jolobe's section accused Gombe of being a witch and claimed he killed Jolobe unjustifiably by witchcraft in order to make Tapita well, and the section left Gombe's village and joined another (Marwick 1965: 268–72).

In this example Jolobe is suspected of witchcraft because of his unsocial behaviour over the child's funeral. Once again we see different parties interpreting misfortune in different ways, influenced by their relationships with the principals involved.

Evidence world-wide provides support for the proposition that antisocial behaviour can draw accusations of witchcraft. We are told from the Tangu that surly, secretive and unsociable men, men who are taciturn or nonconformist, are suspected *ranguma* (Burridge 1965: 231–2). Among the Navaho, unsocial persons can be thought to be witches.

Case V-6

'There was a man near Chinlé – my grandfather's brother. This man was a sly one and people suspected him. He acted queer and was always threatening people. People started to get sick and die. So my grandfather decided to find out what to do to witch. On one moonlight night he followed him.

First he went to the graves of twins who had died and were buried in the rocks near the mouth of the Tsegi'. He took out some of their brain. Then he cut off their finger whorls and toe whorls. After that he went around and robbed the grave of a rich man who had died a few days before.

After that he went up the canyon that is just south of the Tsegi'. My grandfather hid and watched the trail that led up to the cave. He saw a lot of wolves and coyotes with straight tails sneaking up that trail. They disappeared in a hole in the canyon wall.

After a while my grandfather sneaked up the trail. He found the way around a big rock that covered the opening of the cave. Hidden behind a big rock he saw a band of some twenty witches with their wolf and coyote robes off. There was a fire burning in there and he saw that they had made sandpaintings of the rich men in the country and one of the agent at Fort Defiance. With small bows made of the shin bones of dead people that shot beads into the sandpainting while they sang the *'Aadak'anshgii*, Shooting 'ways'. Then they spit, urinated and dunged on the sandpaintings as they sang some songs about *Inzini*, offal.

Some time later my grandfather's brother who was a rich man and a headman died from a strange sickness. My brother (the witch) had been hanging around his camp after the girls – his own female relatives. So my grandfather made up his mind to kill him.

There was a big *'Aanaadjih*. My grandfather took his bow and arrow. When he saw his brother he got behind a horse and called to him, "Come over here, brother. I have a present for you." When the witch got near my grandfather shot an arrow right through his eye as he said, "That's the present I have for witches (Kluckhohn 1967: 208).'"

Very few analyses of witchcraft describe the behaviour by which individuals may come to be regarded as witches. Western Apache witches are motivated by hatred or intense dislike (*kɛdn*). This is their only motive; without it there would be no witches. Basso gives a detailed account, from informants' statements, of how one Apache knows that another hates him. His question, 'How do you know someone hates you?' elicited the following information:

> He refuses you help when you need it.
> He fights you without provocation, out of desire to discredit you.
> He is quick to be angry with you, and expresses his anger both overtly and in hidden, invidious ways.
> He distorts truths about you in order to discredit you, and he tells deliberate lies about you.
> He threatens you, and he informs on you to the Apache police about offences Apaches believe should be settled without recourse to courts of law.
> He tries to make you feel inadequate by making you the butt of pretended jokes that are meant in earnest.[11]
> He steals from you.
> He propositions your wife, and if he can he fornicates with her.
> He may have had quarrels and fights with you in the past, and he hates you because of them (Basso 1969: 42–8).

Case V-7

In August 1966 X developed severe pains at the base of his spine. A medicine man diagnosed them as due to witchcraft and X accused Y of bewitching him. Y was a man of about fifty who belonged to a clan unrelated to X's clan. X argued that Y claimed he had power that helped him find lost objects and win at gambling and told him when to buy horses and pick-up trucks. Consequently he could be a witch. He claimed Y had shown he hated him by accusing him falsely of stealing a saddle; by refusing to give him apples when he was unemployed and his family needed food; by propositioning his wife at a ceremonial; and by

---

11. See Basso 1979 for an analysis of the sociology of Apache jokes.

exaggerating the facts about an incident when X was thrown from his horse, in order to portray him as an incompetent rider. Consequently it was reasonable to assume that Y had bewitched him. Everyone agreed that Y had behaved in this way and that therefore the accusation was plausible. Y did not defend himself, which, although not regarded as proof, is taken by Apaches as close to being an admission of guilt (Basso 1969: 45).

For England, however, Macfarlane argues from the Essex data that witches were not necessarily more antisocial in their general behaviour than other persons. For example, witches were not prominent among women who were regarded as scolds (Macfarlane 1970a: 158–60),[12] and often were frequent churchgoers (Macfarlane 1970a: 188). This contrasts with Keith Thomas' claim that suspects often were antisocial in their general behaviour (Thomas 1973: 261, 633). In fact, a witch might be too *good* a neighbour. The suspect who appeared too solicitous over the welfare of the victim could find the suspicion confirmed (as in the example of Stürmlin in Case II-1) (Macfarlane 1970a: 172). What particularly characterized suspects was begging and reacting strongly when refused. Larner argues that suspects were likely to be quarrelsome and have a reputation for performing supernatural deeds, and could have a reputation for healing as well as for supernatural harm (Larner 1985: 72–3).

In many cultures, any kind of strange and unusual behaviour is grounds for suspicion. To the Tangu, any act that is not normal, preferred, expectable or commonplace is *imbatekas* (Burridge 1965: 230). Among the Navaho, witchcraft was dangerous for the witch as well as for the victim. It made him 'reckless', and witches could be recognized by their foolish behaviour (Kluckhohn 1967: 61). Eccentrics and the mentally ill may come to be regarded as witches because of their unusual behaviour. The following Ceŵa case illustrates this.

Case V-8

Katuule was a member of Gombe's section of his village. He was a classificatory mother's brother of Gombe's first wife, Eledia, and was married to Gombe's eldest sister, Luwa. He had always been eccentric. As a boy he suffered attacks of insanity and had to be tied up to prevent him falling into the fire. He liked performing conjuring tricks, and twice

---

12. *Scold*: 'a troublesome and angry woman who, by her brawling and wrangling amongst her neighbours, doth break the public peace and beget, cherish and increase public discord' (Thomas 1973: 631).

this odd behaviour had been thought to be the antisocial behaviour of a witch, and he had submitted to the poison ordeal and been cleared of suspicion. He was jovial and amiable, and would go off on hunting trips and forget to return for weeks on end. In his old age his history of strange behaviour caught up with him, and he was blamed for the illnesses of his matrikin. People believed he was a 'real *ufiti*', a witch driven by addiction for flesh rather than by malice. A little girl, Neli, told her grandmother, Eledia, that she had seen Katuule stealing from her garden. This angered Katuule. Neli became ill and the diviner said Katuule was trying to kill her with witchcraft. He gave her a protective charm to turn Katuule's witchcraft back upon himself. The diviner claimed that this was happening and Katuule's health was declining. Edisoni, one of Gombe and Eledia's sons, became ill when he was living away at school in Fort Jameson, and had to be brought home. Some villagers said he was suffering from a mystical illness caused by his failure to take 'medicines' after his first nocturnal emission; but most believed it was witchcraft. Eledia sought help from a diviner, who gave Edisoni medicines so that he would be able to identify his mystical attackers. When he used them he heard people talking about him outside his hut, but was unable to identify them. He died, and Eledia claimed Katuule had killed him. She also blamed Katuule for the death in Lusaka of her eldest son, claiming Katuule sent him medicines which killed him.

Because of the hostility against him, Katuule and his wife began living away from the village in a hut in his maize garden. People said it was because he realized they knew he was a witch. It was said that he converted his wife to witchcraft, but did so when she was too old, and in consequence she caught leprosy (Marwick 1965: 273–6).

Marwick estimated that in about 30 per cent of his Ceŵa cases there was no quarrel or tension between victim or accuser and accused. In most of these the accused were persons he considered eccentric (Marwick 1965: 291).

Individuals considered too successful or too lucky may be accused of having gained their success by witchcraft. A proverb from the Bemba of Zambia states '. . . to find one beehive with honey in the woods is luck; to find two is very good luck; to find three is witchcraft' (Gluckman 1965: 59). Kluckhohn records that of 222 cases of Navaho witchcraft accusation that he collected, 52 per cent of the accused (115 cases) were rich, and 63 per cent (140) were ceremonial practitioners (Kluckhohn 1967: 59). (However, some of Kluckhohn's cases are hearsay, and are not reliable as evidence of actual events. Marwick's Ceŵa material is more reliable, as it is based upon more detailed questioning.) We have already commented on the Navaho view that people become witches in order to become rich, and that ceremonial experts may be witches who enter into fee-splitting arrangements to extort fees. Rich Navaho try to avert accusation by behaving generously (Aberle 1966: 49). Many Apaches believe accumulation of wealth requires mystical power to be used at the expense of others (Basso 1969: 54). Persons accused among the Apache are generally among the better-off, and accusers are usually from among the less well-off (Basso 1969: 59).

Functionalist Explanations of Witchcraft

The rich and the powerful are often the target of witchcraft accusations. Those who achieve success arouse resentment and envy, and this may be expressed in accusations against them. They also fear that their wealth and social position will attract witchcraft from rivals and from the envious.[13] Social scientists often examine witchcraft within a wider context than simply the interpersonal relationships between individuals. They look at it in terms of its consequences for the society or the group, and not only for the parties directly involved. When analysing the effect of witchcraft beliefs and accusations on the wealthy, functionalist anthropologists stress their role in promoting the distribution of wealth, maintaining egalitarianism or hierarchy, and inhibiting social change (Gluckman 1965: 59; Kluckhohn 1967: 119–20). They regard these social consequences as reasons *why* witchcraft beliefs exist.

Functionalist explanations are based upon the idea that institutions exist in order to fulfil needs that must be met if a society is to survive. The needs they postulate are broad, general needs basic to all societies, such as the need for conflict resolution and for the socialization of human beings into cultural norms and values. By preventing the accumulation of goods, or maintaining egalitarianism, witchcraft accusations against the rich and ungenerous are maintaining a particular pattern of social structure that is adapted to meet basic social needs.

Functionalists usually consider the fundamental need of a society to be the maintenance of an equilibrium – a stable relationship between the institutions that constitute the society, and hence between the activities that they carry out. Consequently society is a system. It consists of a set of integrated parts and integrated activities that promote a tendency towards equilibrium (Parsons 1951). Stability and equilibrium are natural consequences of the system. They are not, for example, the result of a dominant social group using its power to promote compliance with its interests.

---

13. Of 164 victims of Navaho witchcraft, 133 (81 per cent) were rich (Kluckhohn 1967: 60).

Witch Suspects: Suspicions from Personal Qualities

During the 1940s, 1950s and 1960s many anthropological analyses of witchcraft were made within this kind of analytic framework. Their aim was not only to discover the kinds of relationships involved in witchcraft activities, and particularly in accusations, but also to link these activities to the maintenance of a social equilibrium. Classic examples of this are the analyses by Max Gluckman and his associates (for example, Gluckman 1959: 81–108; Marwick 1952: Turner 1957), which used a sophisticated functionalist model that viewed society as a dynamic balance of forces, in order to argue that conflict and disruption generated within a narrow, close-knit set of social relationships can lead to integration and strengthening of ties over a wider set of relationships. Witchcraft accusations enable the breaking up of personal relationships that, despite having a high cultural value, have become conflict-ridden. Once this has occurred, new, stable relationships can be forged between the wider parties and communities affected by the break-up, and this is fostered by the operation of values common to all the parties. (As we shall see in Chapter 6, this is a contentious claim.)

The equilibrium model gave many insights into the operation of certain kinds of relationships, and is useful for the analysis of certain kinds of problems (Gluckman 1968). Sometimes the distortions in the data being analysed that result from treating phenomena as though they constitute, or are moving towards, an equilibrium may be compensated by the insights that are achieved. However, the assumption of equilibrium tends to lead to playing down negative consequences of action, for persons and for social groups.

One reason for this is that functionalists tend to take for granted the nature of beliefs and the meanings actions have for the members of a society. They regard them as secondary factors, necessary to ensure the fulfilment of basic needs, and the whole field of meanings is made subordinate to the integration of social activity. Peoples' motives are not regarded as the real reasons for their actions. The true reasons, of which they are unaware, are the supposed long-term consequences of their actions for the structure of society. For example, people may take certain actions because they believe in witchcraft, but the functionalist may argue that the true reason is because they need to modify particular relationships, and this is related to some kind of vague social need. To put it another way, witchcraft beliefs exist because a society requires stable social relations, not because people are trying to explain misfortune or to pursue some other interest. Once the problems of functionalism as a general sociological explanation became manifest, students of witchcraft often turned to examining its symbolism and meaning rather than the consequences of accusations for social relationships (Beidelman 1980; Crick 1973; Lienhardt 1951; Needham 1978).

As an interpretation both of social developments and of historical events, functionalism is also gravely deficient. Many societies quite clearly are not in the kind of stable equilibrium that functionalists postulate, and we cannot assume that they are moving towards it. Many kinds of conflicts, including conflicts internal to societies, do not promote this type of equilibrium and stability. Functionalism cannot explain major social conflicts and social changes; but often it cannot adequately explain such social stability as does exist either. It cannot be determined what the specific needs of a social system or society are, and it cannot adequately be demonstrated how an institution fills particular needs (Cohen 1968: 34–68: Rex 1965: 60–77).

And even if this could be done, at the level of generalization required by functionalism the knowledge would not supply us with adequate answers to the kinds of questions most social scientists, and most other persons, wish to ask about the operation of society. These usually concern issues that have some kind of political connotation. In order to examine them we need analyses in terms of intermediate concepts, such as power, politics, interests, and economics, and not of distant concepts such as basic social needs, and we must focus upon some intermediate level of activity, and not upon the distant level of an abstract society. Among the questions we wish to ask about witches are: how are witch-beliefs affected by different kinds of power situation, and what happens to accused persons in these situations, and why are there witch-hunts?

A Conflict Theory approach focuses upon politics, and is concerned particularly with the relationship between beliefs and accusations and the pursuit of power. It also is concerned with the wider consequences of witchcraft beliefs and accusations, which are seen as influencing the form and pattern of change in a society. But this is not regarded as the reason for the existence of the complex of beliefs and actions associated with the idea of witchcraft. This lies in the competitive nature of relationships between individuals and between groups. We shall examine the relationship between witchcraft and power, politics and social change in Chapters 8, 9, 10 and 11.[14]

---

14. For discussion of different types of explanation of witchcraft beliefs and accusations, see Basso 1969: 2–3; Devons and Gluckman 1964: 241–53; Mair 1969: 199–221.

—6—

Witch Suspects: Suspicions Rooted in the Structural Position of the Accused

Strangers and Foreigners

In this chapter I wish to discuss persons whose position in society, relative to that of other persons, is likely to cause them to be thought to harbour the kinds of feelings and attitudes believed to typify the witch. One such category of persons commonly accused of witchcraft is strangers or foreigners. Tangu believe people of uncertain ancestry are *ranguma*. The most dangerous *ranguma* are non-Tangu. They believe that those Tangu who are *ranguma* are generally members of other neighbourhoods (Burridge 1965: 232–3). The word *ranguma* is very similar to *rangama*, the word for stranger, and Tangu play on the similarity between the two terms (Burridge 1965: 249 n.10). Baxter's analysis of witchcraft among nine East African pastoral peoples shows that among several of them there is some relation between witchcraft and foreigners or outsiders. Mandari witches are usually mistrusted immigrants, the descendants of destitute outsiders who were given land and became clients of their land-holding patrons. They are believed to seek illegitimately to take over ownership of this land. Nandi witches are associated with 'interlopers'. Among the Samburu only persons with Turkana ancestry can be witches. Northern Somali witches are often Yibir bondsmen, a scorned group regarded as outsiders. Among the Boran, persons with the evil eye – the power mystically to inflict minor misfortunes – are individuals who have withdrawn from the fully pastoral Boran economy and are seen, and see themselves, as only part Boran (Baxter 1972: 165–9).

Among the Navaho the residents of two areas are the subject of particularly intensive gossip about witchcraft. Both are, in a sense, 'foreign'. One population has a large intermixture with the Pueblo Indians. The other is called 'enemy people', and is descended from Navaho missionized in the eighteenth century who in the past fought with the Spanish against their fellow Navaho (Devons and Gluckman 1964: 248; Locke n.d.: 189, 194–6, 258–9, 301–2, 372). The Western Apache believe unexpected strangers may be carrying witch poison (Basso 1969: 34).

Among the Eastern Shona, 54.5 per cent of the allegations of witchcraft and sorcery collected by Crawford are against a neighbouring stranger (Crawford 1967: 137).[1] European folklore tends to stress the role of outsiders as witches. In practice, however, the great majority of suspects were neighbours of their accusers. Although European beliefs may emphasize the outsider, reality was different. Briggs suggests that this popular over-emphasis on the outsider may be an expression of peoples' reluctance to admit to themselves that evil can be internal to their community (Briggs 1989: 27).

Ambiguous and Marginal Social Positions

Another common category, related to the previous one, is persons in an ambiguous position in the society, who combine the qualities of insider and outsider. They are perceived as members of the group, whilst simultaneously being perceived as members of another group with a strong demand on their allegiance. Consequently they are thought to have conflicting loyalties. They can be seen as the 'enemy within' and the 'traitor within the gate'. This is classic witchcraft language, and their presumed treachery may be believed to take the form of witchcraft directed into the group in which they reside (Mayer 1970: 60–4; Winter 1963). Some of the examples mentioned above probably fall in this category. In his analysis of witchcraft among East African pastoralists, Baxter describes as ambiguous the position of some of the minority groups whose members are thought to be witches (Baxter 1972: 165). The idea of the witch has the characteristic of ambiguity, defining a person who is a member of a human community whilst being morally outside it (Beidelman 1986: 138).

The Zulu local group traditionally has a core of closely related patrilineal kinsmen. When men marry they bring in wives who are the daughters of other local groups and members of other clans. Wives have ties with their natal communities and with their own clans, as well as with their husband's community. We can see that there is an ambiguity in their position that is likely to affect the way in which they are viewed. Zulu women often were accused of witchcraft by members of their husband's community. Gluckman sees them as scapegoats on whom are focused the socially-induced tensions that arise between members of the group but cannot be expressed between them without revealing its disunity (Gluckman 1959: 98–100; Gluckman 1965: 223–4). He states that among

---

1. Crawford's allegations were collected from court records, and relate only to those that reached the courts. Consequently they are not reliable as a measure of allegations made locally.

the Zulu, in contrast to men, women are perceived as having evil in their very nature (Gluckman 1965: 224). Accusation of women married into a patrilineally-based group, by members of that group, appears common. However it seems less common in societies where men marry into their wife's matrilineal group for such men to be accused. It appears that ambiguity on its own may not be enough. It may need to be exacerbated by other factors, such as beliefs about the nature of the sexes, and the pattern of political competition within a group (Crawford 1967: 151–7).

The idea of ambiguity in the social position of particular categories of persons has been applied to other situations in which people are accused of witchcraft. We have seen that the Nupe associated a particularly evil and antisocial form of witchcraft with women. Nadel argues that this is because there is a contradiction between the ideal role and the real position of many women in Nupe society. A woman should be subordinate to her husband, a faithful wife staying at home to look after her husband and children. In fact, trading makes many married women wealthy and enables them to behave in a markedly independent fashion. Nadel suggests this anomaly between ideal and reality causes women as a sex to be associated with witchcraft (Nadel 1954: 174–80).

Because there is a widespread tendency for women to be accused of witchcraft more commonly than men, Gluckman has suggested there is a deeper, more basic ambiguity in the position of women. They bear children, which ensures the continuity of the group and influences its power and social standing. However, when these children grow up they compete for power, property and social position, thus creating conflict and dissension. Consequently women introduce conflicting social processes *into* the group. That this is the cause of the cross-cultural prevalence of female witches is, of course, difficult to prove. However, the hypothesis that there exists a fundamental ambiguity in the social position of women in many types of society may gain some support from the fact that in many cultures the female sex is regarded as intrinsically capable of evil, whereas men must learn how to perform the kinds of antisocial mystical actions of which women are believed to be capable by their 'nature'.[2]

In many cultures the position of the elderly is an ambiguous one. As

---

2. See Gluckman 1965: 224–7, for a discussion of why women are so frequently accused of witchcraft, and the problems of relating gender differences in accusations to any particular form of social structure. On this latter point, Middleton and Winter hypothesize that among indigenous societies, accusations against women are particularly associated with patrilineal societies that possess the 'house property complex' – where men inherit from their fathers (or other patrilineal kinsmen) through their mothers (Middleton and Winter 1963: 15–16). On the question of sex antagonism and female witches, see also Nadel 1954: 172–81.

the elder generation they should be respected, but in real life, because of their age, they may count for little. They may also be anomalous because they are still in the society and yet they are close to death, on the boundary of life and death. Of Kluckhohn's 222 cases of Navaho witchcraft accusations, he tells us that 173 (78 per cent) of the accused were 'old'. All women accused (19 per cent of accusations) were 'old', as were nearly three-quarters of the men (Kluckhohn 1967: 59–60).[3] Their perceived closeness to death is the reason Navaho give for identifying the aged with witches. They say the very old, because they know that they will soon die, may take all sorts of chances to obtain immediate gain – chances younger persons would be hesitant to risk. Kluckhohn believed the Navaho felt that the aged were almost ghosts, and ghosts are linked with witches (Kluckhohn 1967: 59–60). The Hopi believe old witches sacrifice the lives of others in order to remain alive (Simmons 1963: 120, 209). The Western Apache make an intrinsic association of advanced age with mystical power, and assume that all elderly persons are capable of witchcraft (Basso 1969: 53). Shona do not make a clear division between the living and the dead. Very old persons have become like ancestral spirits, and can cause harm if they are slighted. Consequently elderly persons are feared, because they have some of the characteristics of witches, and a fairly high proportion, especially women, are accused of witchcraft (Crawford 1967: 81).

Navaho find the care of old relatives to be a strain, and there is ambivalence towards them on that account also (Kluckhohn 1967: 105). Crawford suggests this may be an important influence on accusations against the elderly among the Shona (Crawford 1967: 156–7).

In some societies the elderly may occupy more of a marginal position, not sharing fully in the benefits and rewards of societal membership. Because it is recognized that they suffer deprivation, persons occupying such positions may be thought to be envious and to harbour resentment against other social groups. Baxter cites a number of marginal groups in East Africa that produce witches: kinless widows and Yibir bondsmen among the Somali; cultivators among the Boran; poor immigrants among the Mandari (Baxter 1972: 165–9). The very poor among the Navaho are believed to feel malevolent because of their deprived position, and it is advisable to treat them with generosity in case they are witches (Aberle 1962: 49).

Some social historians examining the European Witch-Craze of the sixteenth and seventeenth centuries (which will be discussed in Chapter 10) have sought to explain the prevalence of female accused by reference

---

3. The strictures already made concerning the reliability of Kluckhohn's statistics apply here also.

to the social position of women. Christina Larner relates it to women's marginal position as a result of their subordination in the power structure of Early Modern European society. She suggests that women came to be seen as a threat to the patriarchal order of European society during the sixteenth and seventeenth centuries. Like Gluckman's analysis of the Zulu, Larner points out the association of men with sanctity and women with evil in the ideology of Early Modern Europe (Larner 1985: 86–7).[4]

Many of the categories we have described are likely to overlap each other. The aged often are eccentric and antisocial – a mental consequence of age. Their position may be ambiguous *and* marginal, *and* they may act in abnormal and anti-social ways. Many of them are women, as well as aged, and their position may be marginal or ambiguous on that account also. Elderly Navaho often are rich and are ceremonial leaders, and are likely to be cantankerous or eccentric. A number of perceived reasons for suspicion may be present in an individual, strengthening the likelihood of accusation.

Case VI-1

The Hopi Indian Don Talayesva looked back on his life when he was fifty years old and identified six persons as witches in his native pueblo of Old Oraibi. (These were not the only witches in the pueblo; he believed there were others whose identity he did not know, or whom he did not identify.) All were old persons, one woman and five men. Don said that several of these witches had outlived their lifespan and been called by the god Masau'u. They were prolonging their lives by sacrificing their relatives.

Three of the men were members of Don's wife's clan, the Fire clan. Two of these were his wife's mother's brothers, and he claims their identity was revealed by a third male witch, whom he identified as his wife's 'uncle'. (He is supposed to have confessed to Don's wife's mother in remorse for his witchcraft. He was a very old man and wanted to die.)

The old woman was a member of Don's clan, his mother's mother's sister. When Don was young he dreamed that she was a witch. Later she massaged his back when it was painful and it became worse. He thought she had bewitched him, and told his fears to a Hopi doctor, who confirmed to him that she was a witch. Another of the old men is identified as Don's 'grandfather'. A Hopi doctor accused him, in Don's presence, of making Don ill by mystically shooting poison arrows. The old man was flustered

---

4. The role of women as victims of witchcraft prosecutions in Early Modern Europe will be dealt with in Chapter 10.

by the accusation. This convinced Don that it was true, and that the 'grandfather' was trying to hide his guilt. He thought the old man was trying to kill him in order to prolong his own life.

Nathaniel, the sixth witch, was also old. He was mentally unstable, and the Whites said he was insane. It is his peculiar and unsocial behaviour that is given prominence in Don's account, rather than his age. He was chief of the Ahl society, but his instability and the belief that he was a witch did not bar him from carrying out his ceremonial duties. He behaved strangely, teased and threatened children, threw stones at dogs, and tried to be too familiar with women. Popular belief that he was a witch was confirmed when his wife and children died one after the other and he did not bury them properly, but let the Christian missionaries bury them. He was killing his family to prolong his life, and his refusal to bury them was further proof that he was a witch. Don also identified a witch from the Hopi pueblo of Hotavila. He appears to have been an old man, and he was the best doctor in Hotavila. In his desire to renew his sexual prowess he was trapped into joining a society of witches. As a result he lost interest in life, became very depressed, and died. His son was unsympathetic and told him to die because he had become a witch.

But Don himself was rumoured to be a witch by his wife's clan, because of a series of misfortunes in his relationships, in particular the deaths of his children, that were culturally diagnostic of witchcraft. His four children all died very young. (He believed they were killed by witches from the Fire clan.) Aware of the possibility that people might believe him to be a witch, he sought to deflect accusations by accusing first, or by counter-accusing in his turn. When his third child died he accused his wife's relatives and claimed one of them must have been responsible. By publicly doing so he was also seeking to defend himself against a possible charge that he was a witch and was killing his children. (The deaths of his first three children could be used as evidence against him, just as he used similar evidence to confirm to himself that the mentally-deranged Nathaniel was a witch.) His wife's relatives countered by saying he was unlucky with children because he was careless in his ceremonial activities, had extramarital affairs, and had arguments with his wife that worried the babies and made them pine away and die – all examples of unsocial behaviour.

When his fourth child became ill, his wife's mother would have nothing to do with him or his wife. His wife's relatives blamed him for the deaths of his children and said it was because he had intercourse with other women and did not perform his ceremonial duties correctly. They hinted that he was a witch who had taken his children's lives to save his own. When his wife heard she asked him if it was true. He denied it, and brought his wife's mother and his principal slanderer, Sequapa (a Fire clan woman), to his home. He told them he was not a witch and had never been to the witches' secret convention. He reversed their charge by daring them to 'kill him and be done with it' – the customary method of challenging a witch. But the child died. Don's wife was not entirely convinced of his innocence, and Don himself had doubts until his guardian spirit came to him in a dream and told him that he was not a witch and that the child was killed by a member of his wife's clan.

Don needed living children to prove that he was not a witch like Nathaniel, who had killed his wife and children to prolong his life. People did not ask him to be ceremonial father to their children because they said he was unlucky with children. His four dead children were proof to the Fire clan that he was a witch. Then a Fire clan woman gave Don and his wife her sick child to raise, because she seemed to be unlucky with it. This gave Don his chance to prove that he could raise children. The child survived and grew to adulthood.

Don was zealous in confronting the old man Nathaniel and accusing him publicly of witchcraft, but was very afraid of becoming a witch himself. His buttress against this was his belief that his guardian spirit would never abandon him. By the time he was fifty he was living well and earning $20 a month by keeping a diary for an anthropologist. Some members of the community became jealous of his good fortune and started a rumour that he was a witch. He had told a clan uncle of his dream trip to the witches' underworld *kiva* (described in Case III-4), and the uncle told others. A Hopi doctor said Don had been there in reality (i.e. that he was a witch), because two of the songs he described were real Underworld songs. One of Don's clan sisters advised him not to talk about his dream because some people were claiming it proved him a witch. Another woman, when alone with Don, accused him of obtaining payment from a museum for supplying them with dead Hopi bodies – a reference to his relations with anthropologists. Don denied it, and told her he had only one heart (i.e. that he was not a witch). He was no longer afraid he might become a witch, because he knew his guardian spirit would protect him until he died (Simmons 1963: 404–11, and *passim*).

In this example, notice once again how different interpretations of the same event are held and manipulated by persons concerned.

In the overwhelming majority of recorded instances of English accusations of maleficent witchcraft, the accused are women, despite there being no objection in principle to the idea of male witches. Of 270 suspected witches who were tried at the Essex Assizes between 1560 and 1680, only twenty-three (8 per cent) were men (Macfarlane 1970a: 160: Macfarlane 1970b: 84). Of 101 persons known to have been executed on the Home Circuit in the sixteenth and seventeenth centuries as witches, only seven were men (Thomas 1973: 620).[5] Only two of the twenty Pendle witches, James Device and a John Bulcock, were men. I have argued that women frequently are in a marginal or deprived position, and that there are elements in their position that may make it an ambiguous one. These factors are marked in sixteenth- and seventeenth-century English witchcraft (Thomas 1973: 678), where the stereotypical witch was an independent-minded and argumentative woman. Larner says that she was 'a liberated woman who . . . [insisted] on making an issue of it' (Larner 1985: 61, 84; Macfarlane 1970a: 158–60; Macfarlane 1970b: 84, 88, 93–4).

Keith Thomas tells us that sixteenth- and seventeenth-century English writers believed witches came from the lowest ranks of society. In fact, the majority were wives or widows of labourers, and their confessions indicate they lived lives of poverty, impotence, and desperation. Their desire to escape these social and material conditions was believed to lead them to enter into a pact with the Devil (Thomas 1973: 620–5). Old Chattox and Demdike certainly fall into this category. Chattox confessed that the Devil promised she would want for nothing. Old Demdike said he promised to give her anything she wanted. Typically, these promises were not fulfilled. Having made the witch his servant, the Devil reneged on much of his part of the contract. At her trial Chattox bemoaned that Francie, her familiar, had deceived her. He had deserted her and taken away her sight (Peel and Southern 1969: 27–31).

However, from a more detailed examination of the social position of persons charged with witchcraft in Essex, Macfarlane argues that they were not usually from the *poorest* members of the community. Witches were not, for example, common among those receiving parish assistance. Instead they were from the *poorer* section of the community, and Macfarlane's implication is that they were often from families that were going down in social status and becoming poorer. Usually the witch was poorer than her victim or her accuser. These were as likely to be yeomen as labourers. Yeomen form the largest single category of victims. What particularly characterizes the accused is that they went begging, and they had vicious tongues and were likely to curse if they were refused (Larner 1985: 72–3; Macfarlane 1970a: 158–60; Macfarlane 1970b: 85). (Briggs notes that begging is also a pervasive theme in witchcraft cases in French-speaking Early Modern Europe, but that here denials of neighbourliness that result in witchcraft accusations relate to a wider range of relationships than those simply involving charity. He suggests from this that at this time

---

5. The Home Circuit was the southeastern judicial circuit of England, comprising the counties of Sussex, Surrey, Hertfordshire, Kent, and Essex (Macfarlane 1970a: 61–3).

French villages remained more close-knit than English ones (Briggs 1989: 33).)

In the contemporary stereotype, the witch was an old woman. Because of her age, her curse was greatly feared, as it was believed that the longer a witch lived the greater her power became (Macfarlane 1970a: 162, 164). Demdike and Chattox were feared because of their age. In fact, suspects usually were aged between fifty and seventy. This is what one would expect, if suspicions and accusations had to build up over time. They were often a generation older than their victims, who usually were younger adults. If a child was the victim, a parent was usually the person who had aroused the witch's malice. Most witches were married or widowed, and the proportion of widows among those accused was especially high. The Essex evidence suggests just over 40 per cent were widowed. Thus the status of being a woman and poor was compounded by age and marital status, suggesting that age and widowhood created particular tensions in the local community (Macfarlane 1970a; Thomas 1973: 519, 620–2, 671).

Personal Relations of Conflict

Witchcraft is associated with malice, envy and hatred. Consequently, when individuals seek the witch who caused their misfortunes they are likely to look for someone with whom they have quarrelled or are in some type of conflict, because such persons probably hold bad feelings towards them. Witchcraft therefore frequently is associated with interpersonal conflict, and persons often look among people they know for the witch responsible for their ills. This especially is true of societies where a few personal relationships serve most of an individual's needs, and where persons are likely simultaneously to be in several different social relationships with one another. Consequently it is characteristic of small-scale societies and of small, stable local communities. Gluckman points out that in such societies and communities, as these personal relationships are so important to an individual, all the events that affect him tend to be explained in terms of them. People who suffer misfortune believe this is likely to be a consequence of something amiss in their personal relationships (Gluckman 1959: 94; Marwick 1965: 189–90). As we saw in Chapter 4, the operation of witch-finding techniques confirms this belief.

In any society there are some relationships in which the relative positions of the parties bring them into opposition. They generate conflict between the persons involved, often because the demands of the relationship force their interests into conflict. In studies of witchcraft, social anthropologists have tended to focus upon these types of relationships and upon positions of ambiguity, so much so that they often give the impression that these are the only important patterns of witchcraft accusations (Gluckman 1965: 216–26; Mair 1969). This is because social anthropologists have had a particular interest in social structure – in patterns of social relationships. However, for reasons that I will discuss in the next chapter, I believe that other categories of persons – antisocial persons, those in marginal or deprived positions, strangers, witch-finders and other mystical experts – are equally likely to be accused, and are more likely to have the label of 'witch' stick to them.

The most commonly cited example of witchcraft accusations expressing conflict that is generated by the structure of interpersonal relations is the relationship between co-wives in polygynous cultures. Women who share a common husband almost inevitably come to see themselves as competitors for his attentions and for the use and inheritance of his property. They are competing to secure their own interests and those of their children. It is easy for a wife to believe her co-wife is being favoured at her expense. Consequently, in many cultures co-wives may accuse each other of witchcraft (Gluckman 1959: 97; Gluckman 1965: 223).

Case VI-2

As a young man, Gombe, the Cewa village headman, became a Christian and trained as an evangelist. He married Eledia, a young woman from his father's matrilineage who was also a Christian. He became headman of his village section. When a witchcraft accusation led to bad feeling against the village headman and a number of sections withdrew and formed their own village, Gombe became headman of the new village. A village headman needs women to make beer and prepare food for his guests, so Gombe took a second wife, Etelina. As a result, he was discharged from the mission. Eledia was bitter and resentful, and became permanently estranged from her husband. Although a polygynist, Gombe continued otherwise to be a conforming Christian, apart from his strong belief in witchcraft, which he shared with most Cewa Christians.

In her opposition to Gombe and Etelina, Eledia was supported by her matrilineage and by Gombe's sisters, and this enabled her to control her three children according to her own wishes. Tension increased between Eledia and Gombe, and in 1947 Gombe and Etelina discovered a horn, supposedly containing witchcraft medicines, hidden in Etelina's hut. They believed Eledia was making witchcraft against them. Some time later Gombe struck Eledia during a quarrel and her matrilineage relatives took him to court. He produced the horn and accused Eledia of practising witchcraft against him, and the court reprimanded her and he won his case (Marwick 1965: 262–7).

In this example, the usual tensions inherent in polygynous relationships are augmented by the fact that Eledia is a convinced Christian and Gombe had been an evangelist. Both should be opposed to polygyny, so Eledia is particularly aggrieved.

Although Ceŵa claim polygyny produces witchcraft, in fact, among the Ceŵa, the incidences of believed attacks and of accusations involving polygynous relations are low. In Marwick's examination of 101 cases of witchcraft, they were a factor only in 7 per cent of believed attacks and 3 per cent of accusations, in spite of the tensions associated with the relationship. (This discrepancy demonstrates the importance of examining actual cases, instead of relying on general statements from informants, when examining patterns of behaviour.) Marwick explains this low incidence in terms of the weakness of Ceŵa conjugal ties. If tension develops it is easy for husband and wife to part, and they usually do so before it becomes the intense hatred that is associated with witchcraft. In Case VI-2, special circumstances, namely the kinship relationship between Eledia and Gombe, made divorce impossible (Marwick 1965: 291–3).

Ceŵa also claim that the incidence of witchcraft between members of the same matrilineage is high, and here their beliefs accord with the facts. The incidences of believed attacks and of accusations are high. One reason for this is political competition over the office of section or village headman. Ideally it is inherited by the oldest surviving brother of the deceased incumbent, or, failing that, the oldest of his sister's sons. In fact, the succession is open to a degree of competition. In Case V-5, the Ceŵa case involving Gombe and Jolobe was not only an example of unsocial behaviour giving rise to an imputation of witchcraft. It also involved competition between the village head (Gombe) and a section head (Jolobe) who was dissatisfied with his position as Gombe's subordinate. Because of the importance of matrilineal ties in village structure, competition for succession to the village headship, or between village head and section head, is often competition between close matrilineal kinsmen. Within the matrilineage, competition also takes place over the ownership and disposal of property, and over control of women of the matrilineage and their children – which is an important aspect of political competition (Marwick 1965: 294).

Case VI-3

Kabambo, a Ceŵa chief, had no sister's son. It was generally accepted that he would be succeeded by his classificatory sister's son, Kasinda. Kasinda had been given the pre-succession name, but he was not well-liked, and a more distantly related classificatory sister's son, Katete, who had a reputation as an arbitrator in disputes, was regarded as a possible alternative successor. When Kabambo died, Katete claimed Kasinda had killed him with witchcraft. He said Kabambo made a deathbed statement to him that Kasinda had penetrated his protective magic by committing adultery with one of his wives. Many believed the accusation, and Kasinda's succession was blocked for a long time. However, ultimately he succeeded to the office because, of the two contestants, he was the more closely related to the deceased (Marwick 1965: 148).

Kluckhohn argued that Navaho witchcraft accusations are socially adaptive because they are usually between members of different extended families. They enable anxiety to be displaced outside the residential group and on to distant individuals. However, examination of his case material demonstrates that, although much of the gossip about witchcraft may concern distant persons, accusations are often made against those with whom the accuser is closely involved (Devons and Gluckman 1964: 247).

Again, Navaho gossip frequently concerns accusations between blood kin, but accusations are more common between in-laws. It appears that Navaho believe consanguineal kin are likely to bewitch each other, but in their relationships they fear affines more than consanguines. Of 103 accusations between kinsmen recorded by Kluckhohn, 81 are against affines, of which the most common type is a man accusing his wife's father. Of the 22 accusations between consanguines, fourteen of these are against a mother's brother. Clearly these are relationships of tension in Navaho society. The relationship between mother's brother and sister's son is the focus of tension generated by contradictions between co-residential and matrilineal ties (Aberle 1961: 117–18, 169–70; Witherspoon 1983). They are likely to be living in different residential groups, but at the same time a mother's brother has a major controlling role in important issues in the life of his sister's son, such as marriage. In the traditional Navaho extended family, the relationship between a man and his wife's father is a very sensitive one. In economic matters concerning the extended family a son-in-law is expected to carry out the wishes of his father-in-law, who co-ordinates these activities but should avoid appearing to command or supervise his daughter's husband (Aberle 1961: 159–64).

As among all peoples, accusations of witchcraft among the Western Apache are constrained by a number of beliefs. Mystical power is associated with age. (Only one of the accused in Basso's 27 cases was under forty.) A person is immune to witchcraft from within his own phratry.[6] (In Case V-7, Y was about 50 years old; belonged to a clan

---

6. Western Apache clans are grouped into three phratries (Basso 1969: 11–14, 21–5).

unrelated to that of X; and claimed to possess power.) And the suspect should have demonstrated hatred (*kɛdn*) for the victim. Of Basso's 27 cases, only five of the accused were women. This is because women's power is weaker than that of men. But of these five, three were accused by a daughter's husband. The son-in-law:mother-in-law relationship is one of some tension among the Apache, as it is governed by avoidance prescriptions, and a man is obliged to do what his mother-in-law asks him, without reciprocation on her part (Basso 1969: 40–5, 55–8).

Frequently, accusations of witchcraft are not made by the supposed victim. (In most cases involving death the victim cannot do so – unless he names his attacker before dying, or is believed to convey this information after death.) Marwick points out that in order to discover any strained relationships that may have caused an accusation, it is necessary to examine the relations between victim, accused, and accuser. The strain often lies in the relationship between the latter two (Marwick 1965: 283–6). Strain on a relationship may not be due to a simple conflict of interests; it may be a more complex process engendered by conflicting or ambiguous social values. In the Nupe example, cultural values stress a wife's subordination to her husband. At the same time, female entrepreneurial expertise is valued. The simultaneous operation of these two values often produces conflict in the relationship between husband and wife (Gluckman 1965: 225; Nadel 1954: 172–81).

The kinds of social relationships within which conflict may lead to witchcraft accusations are often ones that are supposed to be harmonious, where cultural values demand amity, and where the parties are forced to interact constantly. In fact the relationship may be seen as particularly binding, as is the case with marriage in some societies where witchcraft may be one of the few recognized grounds for divorce. Witchcraft accusations may be one of the few ways to terminate such relationships (Gluckman 1959: 81–108; Gluckman 1965: 216–42; Marwick 1952: 232) – which is not to say that the accusers do not believe their accusations.

Marwick argues that among the Ceŵa witchcraft accusations most frequently are found in relationships where competition is not strongly regulated and where desire for the objective of competition is intense, and where no adequate institutionalized outlets exist for the tension generated by this competition (Marwick 1952: 129). Consequently, witchcraft accusations are rare in Ceŵa marital relationships, because divorce is attained easily, or between members of different lineages, because disputes between them may be resolved through legal proceedings. They are common in intralineage relationships, because competition within the lineage is treated as a lineage matter and is not resolved by recourse to judicial proceedings (Marwick 1952: 218–28; Marwick 1965: 291–4). In similar vein, Swanson has suggested that witchcraft accusations are common in relationships that are important to the parties concerned but where there are no authorities entitled to adjust differences that arise between them (Aberle 1961: 117; Aberle 1966: 203).

Misfortunes, then, are often attributed to witchcraft if they occur after quarrels and disputes. English witch-beliefs placed much emphasis on the curse. If a suspect had threatened the victim, or someone closely associated with him, this was important evidence against her. Chattox warned little Anne Nutter that she would get even with her for laughing at her. She said the same to John Moore when he accused her of turning his ale sour. Anne Nutter subsequently died, as did John Moore's son. When Richard Baldwin threatened Demdike and her daughter that he would 'burn one and hang the other', the old woman retorted that he could hang himself. His daughter took ill, and later died. Innocuous or ambiguous comments might be reinterpreted as threats in the light of subsequent events. When John Nutter threatened to have John Redfearn put out of his house, Redfearn told him he would be in a better mind when he returned. Given Nutter's death within a few days, this apparently harmless and placatory statement may well have been interpreted later as a threat. Where there was already hostility, or when one party had a bad reputation, any comment or action might be considered threatening. The curses of the old and poor were particularly potent, especially if they could be considered justified (Thomas 1973: 599–611, 657). Cursing was always used by the weaker against the stronger, and was the most common charge of antisocial behaviour levelled against suspects.

Because witchcraft was associated with envy and hostility, victims often had a suspect in mind even before the offence occurred. Aware that they had offended someone, they were anticipating personal misfortune (Thomas 1970: 59; Thomas 1973: 652). Out of 460 indictments for witchcraft at the Essex Assizes in Tudor and Stuart times, only 50 involved a victim and a suspect from different villages. Of these, in only five cases were they separated by more than five miles. Macfarlane's analysis of the Essex village of Hatfield Peveral reveals that witch and victim usually lived in the same *part* of the village (Larner 1985: 73–4; Macfarlane 1970a: 168; Macfarlane 1970b: 87). The parties in an accusation, then, were neighbours. There is some evidence of wives sometimes bewitching their husbands; otherwise, in contrast to the situation in many societies, kinship relationships appear to have been of little importance in determining the pattern of accusations (Macfarlane 1970a: 169). As we have seen, in England poorer neighbours tended to be accused of bewitching better-off ones, although the difference in living standards between the two often was slight. Macfarlane argues that those accused were rarely the very poor, as there were formal legal provisions providing for them. Accusations were between the better off (including the better-off poor) and their poorer neighbours (Macfarlane 1970a: 151, 205–6).

Macfarlane and Thomas both argue that the actions believed to have aroused the enmity of the witch and provoked the witchcraft often were denials of charity. Charity provides the link with begging and with differences in social status. The accused had been denied neighbourly assistance, such as gifts or loans of food, money or utensils. Many disputes were promoted by the growing importance of private property, which conflicted with traditional values of neighbourliness. They included disputes over gleaning rights and common land. Witchcraft was sometimes believed to be a retaliation for exclusion from communal activities or celebrations, such as funerals or christenings, or a response to other insults (Macfarlane 1970a: 173–6; Thomas 1973: 658–65; see also Briggs 1989: 35–6, for Early Modern France).

The Pendle witches provide many examples. Several of them were said to have lived by begging, in particular the Devices after the death of John Device. Demdike's threat to Richard Nutter not only followed his refusal to pay what she believed was due to her daughter, Elizabeth, but was made after he had threatened and insulted her in the strongest language. John Law suffered a stroke after he refused to open his pack to find some pins for Alizon Device. Chattox killed one of John Nutter's cows because his son deliberately kicked over the can of milk one of her daughters had begged from his father. James Device said that he killed John Duckworth because he reneged on a promise to give him an old shirt.

Witchcraft accusations in sixteenth- and seventeenth-century England appear particularly to express conflict in relationships between the better-off and the poorer members of the community. At this time, social and legal changes were promoting the transition from a community ethic that stressed charity to the poor to an individualistic personal ethic (Larner 1985: 50–1; Thomas 1970: 63, 72; Thomas 1973: 661–3). Consequently the position of the poor was becoming anomalous, and relationships with the poor and the old were coming under strain (Macfarlane 1970a: 204–206; Thomas 1973: 669–73).

A householder would be subject to two conflicting sets of values. He could legally deny his traditional obligations towards the poor. At the same time, persisting elements of a communal ethic, and the clergy's continuing insistence on charity as a moral responsibility, placed some right on the side of the supplicant for alms. Under these circumstances, any hostility that was provoked by refusal to give alms could be seen as being, to some degree, justified. Hence the strength of the refused supplicant's curse. Thomas argues that in consequence of this, a refusal to give charity left the supplicated with a feeling of guilt. Not only did the accuser assume that the supplicant felt malice because of the rejection; his accusation was

A Deed without a Name

a way of assuaging his guilt for what he believed could be a moral wrong. He projected his guilt on to the 'witch' (Thomas 1970: 63–4; Thomas 1973: 673). In fact, the feelings engendered by such a situation no doubt varied greatly, and I suggest that it is not necessary to hypothesize about guilt. It is enough that the supplicated believe refusal was likely to generate malice.

However, Macfarlane and Thomas go beyond a simple association between witchcraft accusations and a situation of social change. They argue that accusations were a *positive* part of the process of social change, and not merely a reaction to it. They helped to terminate personal relationships and end social obligations, and consequently they assisted the triumph of an individualistic ethic over a traditional, communal one. They did so by making those begging for charity liable to being stigmatized as witches (Macfarlane 1970b: 90–5; Thomas 1973: 676). Thomas suggests this is why elderly women, and especially widows, were so often accused. Their position was adversely affected by the erosion of customary arrangements for dealing with the poor and elderly (Thomas 1970: 64; Thomas 1973: 678–9).

On its own I find this an inadequate explanation for the predominance of women witches, as there must have been many male beggars. (Begging was also common in witchcraft accusations in Early Modern France, but here, although the poor were becoming a social problem, within the village community a major change in ethics was not taking place. Yet women predominated among the accused (Briggs 1989: 31–7, 60).) Macfarlane suggests the predominance is due to the fact that women tended to be more prominent in their resistance to change. As wives and mothers, they were a more co-ordinating element in the structure and operation of village life than were men (Macfarlane 1970a: 161). However we have seen that in many societies the position of women puts them at risk of witchcraft accusation, and that there are likely to be several mutually-reinforcing reasons for this.

Functionalist anthropologists have also regarded witchcraft beliefs and accusations as positive forces within society, but they have done so generally within a model of social stability rather than of social change. They have argued that witchcraft accusations tend to operate to promote a stable society. For example, they argue that witchcraft accusations not only reveal stresses in social relations, they enable conflict-ridden personal relationships to be terminated and new, more stable ones to be established in their place (Gluckman 1959: 81–108; Gluckman 1965: 216–42; Marwick 1952: 232–3). In one of his earlier works on witchcraft, Marwick argued that Ceŵa witchcraft beliefs and accusations 'blast down dilapidated parts of the social structure and clear the rubble in preparation for the development of new (but not novel) ones'. They are 'catalytic to the dissolution of social relations which become redundant' in the process of the development of the matrilineage. By 'new (but not novel)', Marwick is referring to a cyclical development of social relations, which he sees as basic to the structure of a stable society; whereas in their analyses of witchcraft Thomas and Macfarlane are describing a process of the replacement of traditional kinds of relationships with totally new ones.[7]

Mary Douglas is critical of this view of witchcraft, which she terms 'obstetric'. She points out that in Marwick's Cewa cases it is younger men, who are the personnel of Marwick's 'new (but not novel)' social relationships that are to replace the 'redundant' ones, who frequently are accused of witchcraft by the elder, who are resisting the changes and who may win their accusations. She shows that in many cultures witchcraft beliefs and accusations can, when viewed from the point of view of social structural developments, at best be a neutral rather than a positive force. They may serve to stress morality and rupture relations that have become intolerably strained, but they can hardly be classified as 'cathartic', as often they aggravate fears and hostilities and are an obstacle to peaceful co-operation (Douglas 1963: 125–6, 141).

Turner goes further. Witchcraft beliefs attempt to explain the inexplicable in societies with limited technology. As such they do not simply reflect social tensions and conflicts. Once formed, they generate tensions as much as they reflect them (Turner 1964: 315–6). Crawford, writing of the Shona, sees witchcraft accusations as a dynamic force that, once created, sets off a chain of events that can have unforeseen, and often personally tragic, consequences (Crawford 1967: 281, 289). For a woman whose children have died, for example, her already traumatized condition must be exacerbated if she knows that others believe that she has killed them, even where no formal punishment is involved. The personal and psychological effects of such knowledge must have driven many persons deliberately to seek the ordeals that they hoped would prove their innocence.[8]

---

7. However, Marwick suggests that witchcraft may not perform this 'function' everywhere that it is found (Marwick 1952: 232).

8. Dr Rosemary Harris has stressed to me the strength of the negative personal consequences, and pointed out that among the Mbembe of eastern Nigeria it appears that many women volunteered for the poison ordeal as a result of them (personal communication).

Political Competitors

To our categories of witch-suspects we must add a final one, political competitors. These will be discussed in the next two chapters, after we have discussed the process of labelling by which the guilt of suspect individuals comes to be accepted widely within the community. As we see in the Cewa examples, political competitors often are in particular social relationships with each other, and this heightens the possibility of witchcraft accusations occurring between them. However, witchcraft beliefs can become involved in political competition in various ways. Powerful individuals, or the representatives of powerful institutions, may persecute as witches individuals with whom they are not in competition, as part of their attempt to maintain or increase their power. When this happens the accused usually are drawn from certain of the categories described in this and the preceding chapter. This situation is examined in Chapters 8 and 9.

Self-Confessed Witches

A problem we must address is that of self-confessed witches, persons who voluntarily confess to being witches. In the Shona Cases II-4 and III-2, Muhlavo confessed to being a witch and carrying out witch-activities with two other women, Chirunga and Ndawu; and Mazwita, Puna, and Netsayi confessed to being witches and poisoning Mazwita's husband. It appears they did so without being pressured by the community. Researchers, and Shona seeking to explain the women's behaviour in non-witchcraft terms, argue they probably confessed to establish a reputation for themselves and make themselves feared. As women, and particularly as fairly old women, they came from a deprived or marginal group in Shona society, which was itself a subordinated part of racially divided and stratified Rhodesian society. Confessing was an attempt to gain notoriety, and consequently a more important – although personally dangerous – status (Crawford 1967: 44–65).

There are cultures in which self-confessed witches appear to be common, and constitute a relatively large proportion of witches. Such spontaneous confessions have been termed 'introspective witchcraft' (Lewis 1986: 51) or 'introverted witchcraft' (Ruel 1970: 333–50). They are reported from southern Ghana, where again they are made by women. Here witchcraft is involuntary. Because of this, having confessed, a woman is not held responsible for her past witchcraft activities, and attempts are made to cure her of witchcraft. This is a costly process, and Ioan Lewis suggests these confessions are a means whereby women, who as a gender are held to be socially inferior to men, can bring pressure on their husbands to pay them attention and show them concern. It is, he suggests, an indirect means of expressing tensions and bringing about changes in marital relationships. In 'ordinary' witchcraft, which he terms 'extroverted', one party directly accuses the other, and this may be an attempt not only to change the relationship, but to break it off altogether. The accusation is a form of direct attack. Lewis suggests the extrovert form of accusation is often made between equals (for example, between co-wives), and by social superiors against social subordinates (for example, a husband accusing his wife). On the other hand, self-accusations frequently are used by social subordinates to influence relations with their social superiors (for example, a wife in her relations with her husband). In southern Ghana this is usually done by confessing to having harmed a child, usually one's own, involuntarily by witchcraft (Lewis 1986: 51–62).

Such confessions probably are more common in cultures where witchcraft is believed to be involuntary or where one can be made a witch against one's will (Mair 1969: 169–72). Among the Banyang of the Cross River area of western Cameroon almost all instances of witchcraft involve self-accusations. Most illnesses, and almost all deaths due to illness, are believed the consequence of the victim's own witchcraft actions. All persons have a were-animal aspect, which usually operates at night. It may be used to harm others, which is witchcraft; but if it does so, ultimately it is caught by religious societies that guard the community against witchcraft, lying and theft. The 'owner' of the were-animal then becomes sick and will die if he or she does not confess. Consequently sick persons are encouraged to make a confession and appease the society. They are encouraged to make as wide and vague a confession of possible misdeeds as they can, and in doing so they will admit to the *possibility* of owning certain were-animals that *could* have caused specific misfortunes. When the confession is remembered by others, it is in a selective manner that fits the publicly known facts of events in the community and is taken as proof that the confessed is a witch. When people die after making a confession their deaths are linked with parts of their confessions in such a way as to confirm that death is due to their own witchcraft.

This complex of witch-beliefs and associated confessions is found among several peoples of the Cross River area of the Nigeria–Cameroon border (Ruel 1970: 333–50). In eastern Cameroon, confession traditionally is a preliminary to exorcism[9] by native doctors to neutralize witches' occult power and enable them to be reintegrated into the community. Persons confessing witchcraft, many apparently voluntarily

---

9. 'Exorcism' is the term used by Fisiy and Rowlands (1989).

or with a minimum of persuasion, graphically describe their witchcraft as involuntary, as a force that got out of control and they are afraid will do so again unless others bring it back under control for them. Suspects who do not confess are held to be very dangerous witches indeed, and in the past were made to submit to the poison ordeal. However, in Cameroon it is widely believed that anyone who is successful must possess mystical power. When witches confess, they are not merely asking for help. They are claiming to be people of some importance, who possess a power that they are temporarily unable to restrain. It is ironic that in Cameroon today the person who confesses to witchcraft, with the aim of being reincorporated into the community through exorcism, may instead receive a heavy prison sentence and a large fine for practising witchcraft (Fisiy and Rowlands 1989: 78–82; Rowlands and Warnier 1988: 121–5).

—7—

Labelling the Witch

Case VII-1

Another of Professor Norman Cohn's accounts of charges of witchcraft made in the Swiss Canton of Lucerne in the sixteenth century concerns a midwife, Dichtlin, and her daughter, Anna.

The depositions were made in 1502. One man claimed that six or seven years previously Anna brought him stewed apple when he had a chill, and told him that nobody else should eat it. After he ate he was unconscious for two days. Another man said that, ten or twelve years previously, his dying mother swore that Dichtlin was killing her out of envy. She also was a midwife, and people called her out more often than they did Dichtlin. A third man claimed that, eight or nine years before, his pigs got into Dichtlin's garden and she cursed his wife, who became lame shortly afterwards. (In reality, she appears to have made an ambiguous comment.) He had also observed Anna sitting in the river and splashing the water between her legs. She told him that she was fishing, but immediately afterwards there was a heavy downpour of rain.

Another man also suspected Anna of storm-raising. He met her after she had been fishing, and she pointed out some storm clouds and told him that they might cause harm somewhere. Another man saw Anna fishing in the river. A storm blew up when he was returning home and he suspected she had conjured it. Yet another said that a great storm blew up every time Dichtlin and Anna went fishing together. A beggar claimed he saw them in the stream holding something between their legs. That night there was a great hailstorm. Three men had seen the two women fishing. Shortly afterwards there was a thunderstorm.

A man said that when his wife called a different midwife to attend her labour Dichtlin warned her that she would regret it. There was a storm, and lightning destroyed his house. He blamed Dichtlin. Someone reported that an ox that ran up and down the lane in front of the women's house had died the next day. Someone else claimed that Dichtlin's husband, Hans, commented to a man who was planting fir trees that his labour might 'bring him more harm than profit'. The man was healthy. A few days later he died. He was not saying that Hans was responsible for his death – he did not know who was responsible (Cohn 1976: 239–41).

In this example, obviously there was a build-up of incidents, suspicions, and informal accusations over many years, culminating in a formal charge. There developed a widespread belief in the community that the accused were witches. Dichtlin and Anna had been arrested previously on a charge of *maleficia* (causing harm by witchcraft), and subsequently released. The incidents recounted above will have been reinterpreted over time, in the light of subsequent events and of the growing reputations of the two women. The accused are mother and daughter, and it is suggested that Dichtlin's husband might also be a witch. Like witchcraft elsewhere, *maleficium* tends to run in families. Some accusations stem from Dichtlin's competitive relationship with other midwives. However, her occupation may have influenced the accusations in other ways. Historians usually claim that in Europe there was an association between midwifery and witchcraft. The profession of midwife was a lowly one. Midwives often delivered babies that died, so it was feared that they might be witches who deliberately killed children for cannibalistic feasts and rendered their remains down to make unguents that gave them occult powers. The placenta, umbilical cord, and caul could be used to work witchcraft, and they had access to these items. And midwives had 'knowledge' that enabled them to perform their occupations. It was believed that they might offer a newborn child to the Devil, who would steal the souls of unbaptized children and unchurched mothers (Cohn 1976: 249; Forbes 1966: 112–32). In medieval and Early Modern Europe, as in other societies, possession of occult power of any kind was likely to arouse suspicion, as we have seen in another of the Lucerne cases, that of Stürmlin (Case II-1). (However, very recent historical research claims to refute the historians' belief that midwives were especially vulnerable to witchcraft accusation because of their profession (Harley 1990).)

Today, peoples who believe in witchcraft usually say that in the past a witch was killed or driven out, a situation now prohibited by law. However, it appears in many cases that persons suspected of being witches continued to live in the community, or in new, autonomous, groups that had split off from the community (Mair 1969: 139–59) – a situation that continues today in societies that have a belief in witches. I suggest this is because there is always more than one possible explanation for misfortune or death. Members of the community may interpret the same event in different ways and attribute different causes. If there is wide agreement that the cause is witchcraft, there may be different opinions as to who is responsible. When Mr M. died many people believed the Paiute doctor's diagnosis that he had been killed by Mrs Porter, but a significant minority believed that Mrs M. killed her husband. Different Shona *nganga* gave different interpretations of the same individual's illness. Gombe, the Cewa village headman, believed the death of Jolobe was due to his (Gombe's) anti-witch medicine and that Jolobe was a witch. But the members of Jolobe's section believed he was killed by Gombe's witchcraft, and left the village because of it.

Writing of the Apaches of Cibecue community, Basso says that the 'guilt' of an accused can never be proved or disproved. Consequently, an accusation becomes a question of the degree to which other Apaches consider it plausible. This depends on the degree to which the actions of the accused are judged to satisfy criteria for suspicion, some of which are vague and allow a wide variety of interpretations. Consequently Apaches disagree constantly about the plausibility of accusations and the 'guilt' of persons accused. For example, there is considerable ambiguity surrounding the concept of 'power' (*diyt'*), and the question of whether or not an accused possesses power often emerges as a focal point of contention in disputes over an accusation. There is also the question of whether the accused's behaviour constitutes hatred of the accuser (Basso 1969: 41–2, 48).

Case VII-2

In September 1967, X became suddenly ill. He accused Y of practising witchcraft against him. Y was 43 years old and a member of a clan unrelated to that of X. X claimed Y had power, because he could call horses from miles away; his father had been a medicine man and taught him many songs, and when drunk he had boasted of knowing songs for seducing unmarried women; and he had been accused of witchcraft once before. He claimed Y had shown hatred towards him by accusing him of stealing two lassos and by fighting him at a drinking party when X accused him of propositioning married women.

Y challenged the accusation. He admitted to possessing horse power, but said he would never use it against anyone. He pointed out that he had successfully refuted the previous charge of witchcraft. He still believed X had taken his lassos. He said that he joked with women but did not offend them, and did not sleep with any woman other than his wife.

There was much disagreement over the plausibility of the accusation. Eleven persons told Basso they thought Y might be guilty, but they were not sure. Neither party retracted his claim, but X had a curing ceremonial and felt better, and talk about the accusation died down (Basso 1969: 46–7).

Often the interpretation persons place upon an event is influenced by the social and personal relationships they have with the parties involved, with the result in many cases that there is no widespread agreement as to whether a particular individual is a witch.[1] A young Shona woman accused of witchcraft by her husband's relatives may return to her parents' community, where the accusation frequently is not believed. She may be reintegrated into her natal community, and later remarry. Old persons who are accused do not have this option, and they may confess to witchcraft or commit suicide (Crawford 1967: 245).

In the past controversial accusations may have been subjected to trial or ordeal. But if a party was proved not guilty opponents might not believe it, and Gbudwe's disbelief of his mother's guilt suggests that the convicted might be thought innocent. In his analysis of Azande witchcraft, Evans-Pritchard drew attention to the way a closed system of ideas is able to absorb a certain amount of contrary evidence and allow a degree of experimentation and testing (Evans-Pritchard 1976: 1–32).[2] If a verdict is later shown to be false, this can be explained by the ability of witchcraft to interfere with the divining technique and produce an incorrect verdict. It is not taken as evidence that the set of ideas upon which the system rests is at fault.

This situation is exacerbated by the fact that often no open accusation takes place. Once an accusation is openly made, an appeal is made to the moral feelings of the community (Basso 1969: 45; Crawford 1967: 279–81; Douglas 1973: 136–7). No party may believe it in their interest to take such a drastic step. The accused are likely to have friends and relatives who will take their side, and a counter-accusation may result (Douglas 1963: 129; Mair 1969: 139–42). Because of the nature of witchcraft, the position of an accuser often is an ambiguous one. Witchcraft works through envy and hatred. It is widely recognized that people who readily believe themselves to be hated are people who readily hate. Witch and accuser often can look remarkably alike to the community (Beidelman 1986: 157; Mayer 1970: 58–9). (An Apache who fails to substantiate an accusation has demonstrated hatred of the person accused, by distorting the truth and telling lies (Basso 1969: 48).) Accusation often places the accuser in the position of being accused in turn. It is a dangerous thing to do. In many societies where there is an ordeal, accused and accuser must submit to it together – a recognition of the closeness between malice and accusation.

---

1. Briggs makes the point about lack of consensus over suspects with respect to witchcraft in Early Modern France (Briggs 1989: 40).

2. Luhrmann makes the same observation for contemporary occult groups in England (Luhrmann 1989).

Labelling the Witch

The Azande demonstrate different interpretations of an event. When a death is caused by witchcraft, the victim's kin are obliged to seek vengeance and kill the witch responsible. Some of the traditional means are no longer allowed, so today they must rely on vengeance magic obtained from Zande doctors. First they should ascertain from a prince's oracle that the deceased was not a witch and his death due to vengeance magic. Having confirmed his innocence, they obtain vengeance magic and employ it against the suspect. It is employed secretly, and when the 'witch' dies the avengers cease mourning. They should go to the prince's oracle for confirmation that the individual who has died – the person against whom they directed vengeance magic – was the witch that killed their relative and died because of their magic. However, the kinsmen of the newly deceased also consult the oracle to discover the cause of death. They may be informed that death was due to witchcraft, and set about obtaining their own vengeance. A death that to one party is a justified killing by vengeance magic to another may be murder by witchcraft (Evans-Pritchard 1976: 5–7; Mair 1969: 149–50).

In her examination of African witchcraft, Mair argues that the behaviour of those who are accused is not usually interpreted in terms of the fantasy behaviour that makes up their community's idea of the witch. They are not accused of operating at night in covens, having familiars, sacrificing their relatives and committing incest or eating human flesh. She argues that this is because, in reality, witchcraft accusations are about personal feelings and actions, and not about sinister mystical powers. They are about malice and hatred and quarrelling, as it is these that disrupt the life of the small community and are what its members wish to drive out, and not fantastical imaginary actions (Mair 1969: 144, 152). In making this strict separation of behaviour from ideology, she puts herself in the position of arguing that people do not believe their own cultural beliefs! I consider that the factors just discussed provide a more likely explanation of failure to apply the witch-image to an accused. Where there is no widespread agreement that specific individuals are witches there is no universal application to them of the witch-image. Some may believe they behave in accordance with the witch-image because they consider them guilty. Those who think they are innocent do not. (In Case V-5, Gombe and Blahimu appear genuinely to have believed that Jolobe and Ngombe turned themselves into hyenas in order to eat the corpses of their victims. The members of Jolobe's section did not believe this. Instead they accused Gombe of being the witch who killed Jolobe. In the case that concludes this chapter, VII-4, the whole of Mukanza village believed Nyamuwang'a had familiars that she fed with the blood of her victims.) What is true is that persons generally come to have the witch-image applied to them because of their everyday behaviour, and because of events in their lives. Once they have acquired a reputation for witchcraft, then all their actions tend to be interpreted in terms of the witch-image.

In contrast to descriptions and analyses of African data, much of the published material on North American Indian witchcraft suggests a greater agreement in the community over the identity of witches, with the witch-image being applied to the accused. It could be that North American witchcraft is fundamentally dissimilar to African; but the apparent difference probably results from the manner in which the data were collected. Much early American material was collected from informants who were relying on memories of long-ago events, and consequently it was probably filtered by cultural beliefs so that, in the main, only outstanding cases were recalled – ones where there was consensus over the accused's guilt. Or with the passing of time, cases that were recalled probably became dressed in the cultural paraphernalia of what *ought* to have happened. This material often concerned the beginning of a life of enforced acculturation on reservations as a conquered minority, when the tensions and anxieties this created may have resulted in communal searches for scapegoats, producing a wider consensus over the identity of witches. More recent field researches into witchcraft among Native Americans, such as that of Basso (Basso 1969), reveal the degree of disagreement over labelling an individual a witch.

Witchcraft accusations, then, are not random. They are aimed at particular categories of persons. These are usually persons who display certain patterns of behaviour, occupy particular positions in the social structure, or have certain kinds of personal relationship with the accuser or the alleged victim. They have access to suspect knowledge, or are believed to harbour feelings of malevolence, hostility, or envy towards the victims or their kinsmen and associates. If my analysis of the problem of successfully labelling any individual a witch is correct, logically it would appear that the categories on whose members the witch-label is most likely to become fixed are witch-finders; strangers; persons who are in an ambiguous or marginal position with respect to the *whole* community – such as the elderly; and persons who consistently behave antisocially or strangely. These persons stand in the same kind of relationship to most of the members of the community, and on this account a consensus becomes more likely. This applies even to cultures where witch and victim must be in a particular genealogical relationship for the witchcraft to be effective (Forde 1964: 210–33; Leach 1971). Guilt is more likely to be widely accepted if the suspect is a member of one or more of these categories. Some categories are likely to overlap and reinforce one another. For example, the aged may be both marginal and behave eccentrically, and many of the most respected witch-finders may be old.

Thus the witches of a society are likely to come from these categories. Consensus is more difficult to achieve regarding members of the other category – persons in particular kinds of personal relationship to the accuser – unless they also belong to one of the vulnerable categories. They have different relationships with different members of the community, and a variety of interpretations are likely to be made of their behaviour. Anthropological data provide support for this hypothesis. Frequently it is reported that people do not acquire reputations for witchcraft until they have been accused several times. Such reputations attract further accusations. At the end of the day, Mrs Porter was held responsible for the death of Mr M. because the Porter family had such a bad reputation for witchcraft. A Zande would have to be held guilty of two or three murders (or of the murder of an important person) before a prince would permit his execution (Evans-Pritchard 1976: 5). A Lele who had been accused several times gained a reputation for witchcraft that no ordinary oracle could destroy, and welcomed submitting to the poison oracle to attempt to clear his name (Douglas 1963: 133). Among the Western Apache only those persons who have been accused on numerous occasions and have refused to defend themselves or have repeatedly failed to clear themselves are held universally to be witches. Public opinion has become so firmly fixed against them that they can do nothing to change it (Basso 1969: 48).

Case VII-3

'That old woman, Y, she does it. Her father did it too, they say. All these people here know that. They don't want her around, and that's why she lives up there at Running Water Crosses The Trail . . . Before they knew about it, she was pretty rich, but people don't want to help her any more. So she is poor now, and angry with hatred at everybody . . .

Last year, at the time to pick up acorns, some women of her clan were going out to get some. Y went down to where they were getting ready and said it: "You women, hear what I say. Let me go to where you are going with you to pick up acorns." Then when one woman said it: "I am not of your clan and you might get angry and make me sick. You have done it before. Let us go alone, or I will stay behind."

Over there, across the creek, Y's brother has a big cornfield. He used to let Y use it to farm. But one time when she wanted to plant it he said to her: "When you were not full of hatred for people I let you use my field, but I don't want these people here in Cibecue to think you're too friendly with me. They may think you will teach me how to do it. There is good dirt where you live. Make a farm up there."

A Deed without a Name

It has always been that way with Y. But before they knew it about her, these people in Cibecue, her mother's sister said that when her body was buried she would leave Y a farm. Then Y witched that woman named V. After that the people got afraid of her. When her aunt died she didn't give anything to Y. She didn't want to give anything to a witch, even though they were close relatives' (Basso 1969: 50).

Writers on witchcraft in Early Modern Europe have also shown that within the local community the acquisition of a reputation as a witch was a long process, and have revealed the problem of developing a consensus.[3] As with the case of Dichtlin and Anna that began this chapter, the example of the Pendle witches involves a gradual build-up of suspicions and informal accusations over many years. As reputations for witchcraft were created, the community's suspicions generated a self-reinforcing belief in the suspects' guilt. The first death attributed to the witches of Pendle occurred seventeen years before the trials. Several times individuals had been accused by neighbours, long before any formal charge was laid against them. By the time of the trials, the reputations of several of the accused already were well established. From accounts of this case, it is difficult not to conclude that they deliberately cultivated their reputations. Such a gradual build-up of suspicions and accusations is reported to be characteristic of English witchcraft cases. Briggs and Ladurie note that it is characteristic of witchcraft cases in French-speaking Early Modern Europe (Briggs 1989: 7–65; Holmes 1993: 55–6 – see also his accounts of the cases of Jane Wenham (1792), 50–1, Annis Heard and Joan Robinson (1582), 53–4, and Mary Smith, 57–8; Ladurie 1987a: 22–68; Macfarlane 1970a: 110–12).

In small-scale communities, traditionally individuals who acquired such a reputation could be killed or expelled, or were forced to take an ordeal or sought to take it. Or they might remain within the community and be ostracized, but be treated respectfully for fear of giving them offence. On the other hand suspected witches might prove useful. Their power and knowledge were of value to the community, as in the case of Mohave shamans. Suspects themselves might benefit from their reputations. People were afraid of them, and might give them a degree of respect or power. Finally, in some cases witches might repent or be cured.[4] Repentant witches might be a particularly effective defence against witchcraft, as

---

3. See, for example, Briggs 1989: 35–6, 39–42; Ladurie 1987a: 22–68. On the problem of consensus see Ladurie 1987a: 62–8, the case of Marie de Sansarric.

4. For an account of the treatment of suspected witches, mainly African, see Mair 1969: 139–58. For Kaguru witches, see Beidelman 1986: 138–59.

with the *lelú* of the Nupe and the Tangu *ranguma*-killer and among the Nyakyusa. Crawford records that Shona are aware that a person accused of witchcraft, even successfully accused, often fails to acquire a reputation as a witch, and rationalize this by the belief that witches may be cured of their witchcraft (Crawford 1967: 271–6).

In Early Modern England, a reputation for witchcraft might be an old woman's last line of defence against mistreatment by her neighbours. Fear of her powers might cause them to treat her carefully and behave charitably towards her (Holmes 1993: 52–4, 57; Larner 1985: 85, 90–1; Thomas 1973: 674–6),[5] just as fear of offending unknown witches might cause people to behave sociably in their everyday lives (Macfarlane 1970b: 93; Thomas 1973: 675–6). Chattox, Demdike, and others of the Pendle witches appear to have cultivated their reputations as a lever in their relations with their neighbours. But the position of such persons was a dangerous one. There was the constant possibility of violent informal action against them or, as happened in the case of the Pendle witches, of suspicions and accusations building up until a formal accusation was made. It is clear that some of those stigmatized as witches believed in their supposed powers, at least for beneficent magic.[6] Well into the nineteenth century, in English rural communities there were individuals with reputations for maleficent witchcraft, as well as persons believed to have powers for curing and other benevolent purposes.

If we pursue our hypotheses concerning the social categories whose members are most likely to attract the witch-label, we may be able to comment upon the witch in Late Medieval and Early Modern Europe and upon contemporary witch-surrogates. Labelling explains the predominance of women, the elderly, and the poor among European witch suspects. In contemporary industrial societies the social, economic and technological changes that have accompanied urbanization and industrialization mean that individuals in strained relationships no longer have to interact. Often they can move away from each other. But there remain the marginalized, strangers, the 'strangers within' whose loyalties are suspect, and persons who are considered 'odd'. To them are attributed unsocial, malicious, and evil actions and intentions. They supplied most of the witches of Early Modern Europe, and they remain the scapegoats of contemporary society. Minorities are stereotyped as responsible for various social ills, from unemployment to the spread of AIDS. The strange

5. Briggs 1989: 64, says the same of witches in French-speaking Europe in the Early Modern period.
6. For Early Modern France, Ladurie gives a précis of Robert Muchembled's discussion of the psychological processes by which an accused may come to believe that she is a witch (Ladurie 1987a: 18).

and odd are a focus for suspicion, and the 'stranger' remains, in some often unspecified way, dangerous. All this will be discussed further in our final chapter.

A Ndembu Case Study

I close this chapter with an example of witchcraft accusations collected by Victor Turner from the Ndembu of Zambia. The detail gives us an excellent account of the labelling of witch-suspects in this society. The case concerns the involvement of witchcraft in politics, which is further developed in the following chapter.

As the Ceŵa examples demonstrate, there is often a relationship between witchcraft accusations and the pursuit of power, influence and social standing. Many Ceŵa witchcraft accusations were related to competition over the position of village headman, the highest position of authority and influence that a Ceŵa man usually could hope to attain. Anthropologists working in Central Africa developed the analysis of the relationship between accusations and competition for village headship in this region, in particular Victor Turner in his study of the Ndembu (Turner 1957). Turner relates the pattern of competition and accusation to a hypothesis fruitfully developed by Max Gluckman, that competition for social position often is influenced and exacerbated by the simultaneous operation of contradictory principles of social structure (Gluckman 1959: 98–100; Gluckman 1965: 224–5). Turner sees Ndembu society (as he would see all societies) as built upon contradictions between the crucial principles that govern the structure of the community (Turner 1957: xvi).

The most radical incompatibility is between rules of descent and rules of residence (Turner 1957: xix). Ndembu descent is matrilineal, and the ideal of residence is virilocal. The most important ties and obligations should be between persons related by descent through females, but on marriage women go to reside with their husbands. Consequently men live with their brothers, but are separated from their sisters' sons, who will become their heirs (Turner 1957: xix). Men manipulate these contradictory principles in order to build up a following that will support them in political competition, trying to retain control over their wives and children whilst seeking to control their sisters and sisters' children (Turner 1957: xxiii–78). This is possible because there is no shortage of land, and so inheritance of land is not an important factor in social relationships (Turner 1957: 76). Ultimately matrilineal descent tends to be the stronger factor, because of the value placed on matrilineal ties and their association with succession to the office of village headman. As he becomes older, a married man usually chooses to join a village in which he has a claim to succeed to the office of village head, through kinship ties through his mother.

The Ndembu are farmers and hunters living in northwestern Zambia, and were studied by Turner in the 1950s. They live in small villages, with an average membership of about 22 persons (Turner 1957: 34–60). One reason for the small size is competition for the position of village headman. A large village with a population of about 50 persons is unstable and ripe for division, as unsuccessful competitors secede and attempt to found their own villages. In this century division has become more frequent, as men establish farms and independent residences away from kinsmen in order to avoid claims on the income they derive from cash crops (Mair 1969: 130). In spite of the limited authority of headship and the tiny community to which it applies, men are ambitious for the position and the influence and status that it gives, and women are ambitious for their sons. Turner describes a man who lacked such ambition. He was 'a quiet and self-contained person, a devoted husband and father, but without that stubborn individualism, barely concealed by the veneer of politeness and sociability, which is typical of Ndembu of both sexes' (Turner 1957: 101). When the headman of his village died he had not put himself forward as a successor, even though he had the correct genealogical qualifications. Consequently, his fellow villagers despised him, and he left and went to live in his wife's village, claiming his kinsmen were practising witchcraft against him (Turner 1957: 101–2).

A man wishing to found his own village needs women of his matrilineage if he is to be reasonably sure that it will continue and perpetuate his name. As the village develops it becomes a small matrilineage (Turner 1957: 76). Once it contains three generations it begins to manifest oppositions along internal lineage lines (Turner 1957: 79), and conflicts between the branches of the lineage[7] are added to other conflicts within the community. Succession to the headship tends to pass between brothers before being transmitted to the next generation, but, as among the Cewa, succession often promotes competition among persons with genealogical qualifications for the office. A sister's son who is able and ambitious frequently becomes impatient for the office from which he is barred by the operation of adelphic succession, and may seek to secede from the village with his siblings and their children, particularly his sisters and their children, and form his own village (Turner 1957: 87). However, a brother or a sister's son of a current headman who believes he has a reasonable chance of succeeding to the position is disposed to remain. Consequently villages consist of a loose association of free and

---

7. Such lineage branches, which manifest considerable autonomy and a strong group identity, are termed lineage *segments* by anthropologists.

independent elders who are constrained by common interest and strength of personal ties to live together, and not by economic exigencies or political edicts from above. Membership of a long-established village may be buttressed by historical pride (Turner 1957: 104); but once a village achieves a relatively large size and has continued through several generations the conflicts within it lead to its division. This operates through, and is justified by, accusations of witchcraft and sorcery.

Because Ndembu believe in witchcraft, accusations are not necessarily merely cynical manipulations of events and ideas in order to discredit a rival. However, they recognize that politics can influence accusations. Turner mentions a local song concerning two contenders for the position of village head. They consulted different diviners, and each obtained a verdict against his rival for bewitching the previous headman. Each then accused the other. A dispute ensued, and the village divided to form two new villages, and the name of the old village died. Turner says Ndembu often are sceptical of divinations made by others; but it appears that the suspicion they have of political manipulation by others they rarely apply to their own actions (Turner 1957: 143–4).

Case VII-4

Turner analyses the relationship between political competition and witchcraft accusations by a detailed examination of conflicts in a small Ndembu village, Mukanza, between 1947 and 1953. The village was founded early this century, and Kaheli was its sixth headman. It had developed from a bilateral extended family into a small matrilineage, and by the late 1940s the lineage had developed two well-defined branches, Nyachintang'a, genealogically the senior, and Malabu. Kaheli was the senior man of Nyachintang'a (Turner 1957: 99).[8] The following witchcraft suspicions and accusations were influenced by the growing opposition between these two branches.

In 1947 Kaheli's sister's son, Sandombu, then about fifty years old, trapped a duiker antelope and did not give Kaheli his rightful share. Kaheli claimed this showed Sandombu despised him. A few days later Kaheli snared a bush buck and sent it back to the village, where Sandombu took it and divided it, retaining for himself the head and a front leg. When Kaheli returned to the village he asked Sandombu's wife to prepare him some food, but she was insolent and Sandombu and Kaheli had a fierce argument, each threatening the other with 'medicine' (a euphemism for

---

8. Mair gives a precis of the following incidents in an account geared to the relationship between witchcraft accusations and lineage fission (Mair 1969: 128–38).

witchcraft).⁹ Sandombu left making threats, and it was believed that he was going to get a witch to shoot Kaheli's familiar. (The implication is that *both* Kaheli and Sandombu were witches.) Shortly thereafter Kaheli died. No divination was held into his death, for fear of government prosecution, but it was believed in Mukanza that Sandombu killed Kaheli. However, he was not expelled from the village, because without divination there was no proof of guilt.

Sandombu was known to desire the village headship. He was a generous man and a diligent worker, and was seeking to build up a personal following. But his ambition was too naked, and it was believed he would stop at nothing to achieve it. He did not join Mukanza village until long after it was established, and as a consequence he was to some degree an outsider. He lacked close relatives who could form the nucleus of any future village that he might head. He had only one child, rumoured not to be his, and Ndembu often regard sterile men as witches. Hot-tempered and prone to unsocial outbursts, he tended to be erratic in his behaviour. He was a tireless gardener, but this was also taken as evidence of his witchcraft. It was said that he had an invisible familiar that worked with him and was fed with the blood of his victims. Given his other qualities, his success was interpreted as proving his witchcraft. He was an atypical villager, an outsider and a marginal man in Mukanza village – an ideal position to draw suspicion of witchcraft (Turner 1957: 109).

But he was also an economic asset to the community. His seasonal job as a *capitao* in charge of road maintenance enabled him to place villagers in regular employment, and from his income he helped pay for schooling for village boys. (All this was part of his attempt to build up a following and present himself as a generous man.) Consequently he was not expelled from the village, but his reputation was sufficient to bar him from succeeding to the headship. Mukanza Kabinda (aged 60) succeeded Kaheli, with village approval.

Kasonda, 35 years old, was Sandombu's main political rival. He had lived for ten years in towns, and had some mission school education. He was skilled in advocacy and the judgement of cases, an accomplishment valued by Ndembu, and spoke with facility and assurance to European officials and missionaries. Although he behaved towards Sandombu in a friendly manner, he hated him as his rival and feared him as a witch. More than anyone else, he was responsible for circulating the story that Sandombu bewitched Kaheli. He was regarded as too young to succeed Kaheli; but he approved of Mukanza's appointment, as he was not expected to live long. If he lived for about ten years Kasonda then would

---

9. Turner uses the term 'sorcery', but entitles these incidents 'The bewitching of Kaheli Chandenda by his nephew Sandombu'.

be in a position to press his own claim.

Sandombu divorced his wife and married Mukanza's daughter, Zuliyana. He probably hoped this would strengthen his ties with the village and that she would give him children. After a year, when she failed to become pregnant, he began beating her. At the end of 1948, when he beat her during an argument, she called on her relatives for help. Sandombu accused them of interfering in his marriage, and said that the people of Mukanza would pay for it. He is alleged to have left the village claiming someone would die the following day. Next day Kasonda's mother, Nyamwaha, died. As she was dying she claimed Sandombu had killed her. He was called to account before the senior men of Mukanza and told that he had cursed the village. This he denied, but admitted he had been angry. His genuine grief at Nyamwaha's death caused public opinion to waver over his guilt. He was told that he must leave the village until he showed that he could live properly with his kin. After a year he humbly asked for permission to return and was allowed back. In the meantime divination had fixed blame for Nyamwaha's death upon her daughter's husband, who lived in another village. It was said he was angered because Nyamwaha wanted to keep his daughter, her granddaughter, in Mukanza village.

Turner now came on the scene as an anthropological researcher, and employed Kasonda as his cook and general assistant. Because of this, when Mukanza's brother, Kanyombu, died, Kasonda was unable to attend his funeral. (He told Turner that Kanyombu probably was killed by Sandombu.) Then Kajata, Kasonda's 'mother's brother' died. When Kasonda and Turner attended the funeral they found that Mukanza, Sakazao, and a young woman, Ikubi, were all ill, in what Turner thought was a malaria epidemic. The villagers regarded Kasonda as the witch who had killed Kanyombu and Kajata and was bewitching Mukanza. He was accused by Mukanza's wives, who said the fact that he was building a Kimberly-brick house was proof he was planning to succeed Mukanza. Kasonda denied it. (Personal factors also were involved in the accusation. There was jealousy over Kasonda's job with Turner, and Mukanza's senior wife disliked him because he had married her favourite daughter and treated her so badly that she left him. Also, the senior wife was from Malabu lineage, and was afraid she might be accused herself, because that lineage had come to be associated with witchcraft (Turner 1957: 145).)

However, Kasonda was able to convince the villagers that Kanyombu's ghost was causing the illnesses because he was angry with Mukanza, who had omitted to carry out his obligations towards Kanyombu when he was dying. In this way, Kasonda was able to reverse the situation and present Mukanza and other elders as the persons who caused the epidemic.

However, he believed that his ambition for the headship was now compromised by suspicion of witchcraft, and determined that he would establish his own village when Mukanza died. Throughout these events no accusations had been made against Sandombu, despite Kasonda's suspicions. Turner believed the most important factor influencing accusations in this case was a recent increase in the size of Malabu lineage. This had provoked tension between its members and Nyachintang'a lineage, which was expressed in the claims and counter-claims concerning the illnesses and the recent deaths. Mukanza village was developing into two increasingly independent lineages, and this was expressed in matters involving such things as bridewealth and the distribution of meat.

Mukanza and Sakazao recovered, but the young woman, Ikubi, died. Her father accused an old woman, Nyamuwang'a, of having killed her with her witch familiars. It was said she became angry when Ikubi refused to give her some meat. Nyamuwang'a had a bad reputation for witchcraft. She was only half Ndembu, and had spent many years in Angola at her father's village. Ndembu who have mixed tribal ancestry, or who are raised among other tribes, are frequently regarded as witches. Turner says it is because they do not 'quite belong' to the local society, and consequently make useful scapegoats for misfortune (Turner 1957: 151). Like Sandombu, Nyamuwang'a was a very hard worker and her gardens gave a high yield. Her strength and luck were said to be due to supernatural powers. When she was young she had a reputation for nymphomania. Ndembu condemn excessive sexual desire. It causes social disturbance, and they associate it with a propensity towards witchcraft. As she grew older Nyamuwang'a came to be labelled a witch, and a series of deaths were attributed to her witchcraft.

Ndembu often consider old women to be witches (Turner 1957: 151–2). If, like Nyamuwang'a, they are widows or divorcees without resident sons or brothers, they have no one to obtain meat for them and must rely on the uncertain generosity of remoter kin. They constantly grumble for meat, and as witches are necrophagous this is regarded with suspicion. It is significant that the alleged reason for Nyamuwang'a's grudge against Ikubi involved meat. Witches usually attack close maternal kin on whom they are dependent and who neglect their obligations, and old women are usually neglected as they are an economic burden to the community (Turner 1957: 152).

When Ikubi died, the accumulated dislike and suspicion of Nyamuwang'a erupted and she was driven from the village, but Sandombu said she had been accused without proof, and offered her a hut on his farm. In 1952 Nyamuwang'a joined Sandombu, bringing her daughter, Ikayi, who had just returned from the Congo with her baby daughter. Sandombu had used the opportunity to add to his personal following, and now had the nucleus of a three-generation matrilineal group that could be expected to endure over time. By allowing Nyamuwang'a to reside on his farm, Sandombu was openly defying Mukanza village and acting like a village headman, as though he intended to split Mukanza village. His following now consisted of the children of his senior wife and some of his more remote kinsmen from Malabu lineage. They included social misfits. Nyamuwang'a was reputed a witch and her daughter prostituted herself, and Sandombu soon added a wayward young lad who acquired a reputation for sorcery. Mukanza villagers referred to Sandombu's following as 'that village of witchcraft'; but he continued to be involved in the daily affairs of Mukanza village.

In July 1953 came news of the death of a kinswoman living in Shika village. The women of Mukanza sang angrily that it was death by witchcraft-familiars, and faced Sandombu's farm as they did so. Nyakalusa, Sakazao's slave wife, accused Sandombu of harbouring a witch (Nyamuwang'a). As an outsider she could speak openly what was on the mind of all (Turner 1957: 269). Sandombu denied there was any proof that Nyamuwang'a was a witch.

Then came a rumour that headman Shika intended to bring an action for witchcraft against unspecified members of Mukanza village in a chief's private court. It was believed that this was in order to avoid Shika paying damages to the deceased's kin for failing to protect her from harm. This threat to the village restored its unity immediately. In the face of external opposition, internal conflicts were papered over temporarily, and the community presented a united face to the outside. It did not simply ignore the witchcraft suspicions within it. It implied that they were unimportant. Nyamuwang'a and Sandombu were chosen to represent the village in its confrontation with Shika.

By 1953 Mukanza Kabinda was becoming very old and infirm, and the question of succession to the village headship was re-asserting itself. Kasonda told Turner that he was now unpopular with Malabu lineage, and so intended to start his own village. He said that when Mukanza died he expected the village to break up. Some of the villagers would go with him, others with Sandombu, and the rest with Sakazao, the senior man of Malabu lineage. (For all these cases see Turner 1957: 95–111, 118–25, 138–68.)

In this extended example we see the build-up of accusations against two individuals who come to be labelled as witches. Both belong to social categories discussed in Chapters 5 and 6. There is some ambivalence regarding the guilt of Sandombu, the hot-tempered outsider. Nyamuwang'a, the kinless old woman with a dubious past who grumbles for meat, is universally labelled by the community. Once they have been

Labelling the Witch

labelled, their positive qualities of diligence and hard work are reinterpreted as evidence of their witchcraft.

In the process of political competition attempts are made to label Sandombu and his rival Kasonda with the stigma of witchcraft. Kasonda does not have Sandombu's social disadvantages and is astute enough to divert suspicion away from himself, but as a consequence of being accused he determines on a future political strategy of breaking up the present village in order to create a new one of his own.

—8—

Witchcraft, Power and Wealth

Witchcraft against Kings and Powerful Personages

We often find that the personal consequences of being suspected of witchcraft, and even of being successfully accused, are not particularly severe. Witchcraft is treated as an offence against individuals and small groups of kinsmen. It is their concern, and not a matter for the community unless the accused has been labelled and is regarded as a public menace, or is accused of an offence affecting the whole community. Ndembu say of the past that old women witches were burned to death (Turner 1957: 152). Presumably these were individuals to whom the witch label had come to stick and who were held responsible for a number of deaths. In most cases, attempts to resolve witchcraft accusations seek to reconcile the parties involved, or if that is not possible, to enable the relationship between them to be modified or severed. One implication of some of those anthropological theories that argue that witchcraft accusations serve to sever relationships that have become over-conflictual, or to 'blast down dilapidated parts of the social structure and clear the rubble in preparation for . . . new . . . ones', is that the long-term consequences of accusation often are not drastic for the accused (Gluckman 1959: 81–108; Gluckman 1965: 216–26; Marwick 1952: 232).

However, where witchcraft accusations are associated with positions of real power and authority and with statuses of high social standing, elevated well above most persons in the society, the consequences of an assumption of guilt are likely to be far more drastic. In indigenous African kingdoms (Vansina 1962), witchcraft between commoners could be regarded as a matter for kinsmen or as an internal matter for the local group and could be treated relatively leniently. But witchcraft against the king was a different affair. It was always a matter of State, and, at least potentially, a capital offence. If the accused was judged guilty, reconciliation or adjustment of the relationship between the parties was not an option, or was at the mercy of the king. This was due to the high

Witchcraft, Power and Wealth

social and ritual status of the kingship.[1] Emphasis was on punishment for the guilty, even if at the king's mercy. A suspicion of witchcraft against political authority could not be ignored. Accusations must be made, and the accused tried.

In Azande kingdoms witchcraft between commoners was essentially a private matter. When it had caused any misfortune other than death, emphasis was on reconciliation between the parties concerned. Only when a witch had killed several persons would a prince permit his being killed. But witchcraft against the king was an offence for which those found guilty by the king's oracle could only be spared at the king's pleasure. It is said of the Azande king Gbudwe[2] that when he did not wish the death of someone who was bewitching him he would summon the witch to his court, give him the fowl's wing, and tell him to blow water over it. Then he swore an oath by the man's side that no one should kill him. But if he said nothing the witch was dragged from the court and hacked to death with ceremonial knives (Evans-Pritchard 1971: 181–2).

Azande kings might order the deaths of witches responsible for public calamities, such as defeat in war. Kings did not make war without consulting the poison oracle to discover if witchcraft would cause casualties among their military companies (Evans-Pritchard 1971: 255).

Case VIII-1

Between 1885 and 1887 the Azande kingdoms of Wando and Gbudwe were at war. The forces of Wando advanced into Gbudwe's kingdom, but were defeated, and many of King Wando's subjects were killed. Widows wailed, and King Wando told the people to consult the poison oracle and put to death the witches who had attacked his subjects in time of war and caused their deaths. Many persons were killed in this way. Only those among the guilty whom Wando wished to spare were allowed to pay compensation to the relatives of those whose deaths they had caused.

Earlier in this war, when Gbudwe's forces were defeated, King Gbudwe drove from his country several men whose witchcraft was divined to have caused the defeat (Evans-Pritchard 1971: 348–51).

African kings represented their subjects and had a mystical relationship with their welfare (Vansina 1962). An offence against the king was an offence against the community. Where authorities have real power and high status, witchcraft acts that are private wrongs in other contexts

---

1. Functionalists would argue that it is because of the key position of the monarchy in the social structure.
2. Gbudwe was killed by British forces in 1905.

A Deed without a Name

become public offences. At the same time, witches may be blamed for communal evils, and not only for personal ones. In Case VIII-1 witches were held responsible for defeat in war, an offence against public policy. Under such circumstances the process of labelling the witch may be much shortened, or may be absent altogether.

From African Kingdoms we have some evidence of powerful individuals and authorities cynically using witchcraft accusations to dispose of political competitors. The nineteenth-century Ndebele kings, Mzilikazi and Lobengula, whose kingdom in Zimbabwe incorporated many of the western Shona as a subordinate element, are notorious for having used witchcraft allegations to destroy persons they considered politically dangerous, although the number of those killed in this manner may have been relatively small (Crawford 1967: 252, 290). Witchcraft accusations may also have been used in this way, at least occasionally, in Shona chieftaincies (Crawford 1967: 290).

Certainly the Shona demonstrate the interest taken by political authorities in witchcraft accusations. Chiefs and headmen are closely involved in organizing divinations to discover the cause of misfortune (Crawford 1967: 194). Shona divination also illustrates the influence of power on the laying of blame for misfortune. We are told that if the divining dice show a positive response to the name of a powerful person, a chief or a village head, the diviner will state that it is not he but a member of his family who is the witch (Crawford 1967: 192–4).[3]

We do not have to restrict ourselves to highly organized states to see the importance that is given to witchcraft fears and accusations when they are associated with power and status, or to societies in which power and social standing are attributes of formal offices. The Tiv of Central Nigeria occupy a rolling plain on either side of the Benue river (Bohannan, L. 1958; Bohannan and Bohannan 1953; Bohannan, P. 1958). In the 1950s they numbered about 800,000, and in precolonial times, which effectively lasted until near the end of the second decade of the twentieth century, they were an independent society with its own political identity and a common political structure.

Tiv recognized only four relations of legitimate authority. Heads of compounds had authority over the members of their compound. Fathers had authority over their children. Husbands had authority over their wives. And individuals recognized as the owners of particular markets

---

3. Dr Rosemary Harris has drawn to my attention the case of the Mbembe, where chiefs were believed to be witches and sorcerers. When Dr Harris asked if anyone ever accused a chief, she was told that no one would dare. She was told that all ordeals were under political control and could take place only with the chief's permission. (Personal communication.)

had authority over persons attending their market. The Tiv polity was organized on a segmentary basis using the idiom of patrilineal descent. All Tiv were members of a common lineage, and political relationships between communities were perceived mainly in terms of relationships between the branches of this lineage (Bohannan, L. 1958). However, the operation of the political system necessitated the existence of local leaders who were men of considerable power and influence (Bohannan and Bohannan 1953: 33–7). But since authority was accorded only to persons in the four statuses mentioned above, their power was regarded as illegitimate (Bohannan and Bohannan 1953: 91).

Power and leadership were viewed in terms of a physical substance, *tsav*, that some persons had within their bodies and that was the basis for any personal talent, power, luck or wealth (Bohannan, L. 1958: 54–5; Bohannan and Bohannan 1953: 84–5; Bohannan, P. 1958). It was a mystical substance that could be used for good or evil, whose power could be increased illegitimately by eating human flesh (that is to say, through witchcraft).[4] All influential people had *tsav*. They were believed to form an organization, the *mbatsav* (Bohannan and Bohannan 1953: 90–2; Bohannan, P. 1958: 5), that met at night to carry out mystical activities to benefit the land and the community. But they might also meet to carry out cannibalistic orgies, for their personal selfish ends.

Because of its nature, Tiv were ambivalent towards *tsav*, and hence towards those who possessed it. On the one hand the *mbatsav* were greatly feared. On the other, as possessors of *tsav* they provided the only protection against persons using their *tsav* for unsocial purposes. Influential and powerful men in one's local community were necessary to provide protection against outsiders and to give the community prestige. But they were hated and feared as witches who had gained their power through witchcraft, killing their close agnates to increase their *tsav* and to repay blood debts to more distantly related witches (Bohannan and Bohannan 1953: 35). Consequently their persons, their property, and their personal qualities were the target of mystical and physical attack.

When, with the advent of colonial times, political leaders were appointed by the colonial administration and their positions formalized as political offices, Tiv beliefs about the nature of personal influence and the illegitimacy of power promoted periodic witch-hunts that attempted to destroy these appointed officials. However, it appears that such witch-hunts were not simply a colonial phenomenon. Although Tiv were not aware of it, they were a recurrent phenomenon generated by the operation

---

4. 'Consuming human flesh' was a mystical, not a literal act, although the colonial authorities misinterpreted the belief to mean cannibalism.

of the traditional political process.[5] Once influential men consolidated themselves in power, Tiv ideas about influence and power promoted the development of indigenous movements to destroy them – to destroy the *mbatsav* who had turned to evil and selfish activities; and these movements operated through the idiom of witchcraft beliefs. The powerful were reduced through witchcraft accusations. As in other situations involving power and wealth, witches were believed to be causing not only personal, but also communal misfortunes – the *mbatsav* were 'spoiling the land' (Bohannan, Paul 1958: 8).

Witch-Hunts in Indigenous Societies

Case VIII-2

An example of the use of witchcraft accusations for political purposes comes from the late nineteenth-century Navaho. In 1879, eleven years after the Navaho were allowed to return to their homeland from confinement near Fort Sumner, Navaho raids were increasing against the Zuni and against New Mexico ranchers. The two most powerful leaders, Manuelito and Ganado Mucho, became worried that this would bring the Navaho into serious conflict with the Whites. They spread the rumour that witchcraft was to blame for the raiding. Then they drew up a list of those persons involved and sent out men to kill them as witches. In this way over forty raiders and suspect witches were killed, and the raiding ceased (Aberle 1966: 50; Locke n.d.: 402–3; Terrell 1972: 242–3).

Today, this story is invariably presented as an account of a cynical manipulation of witchcraft beliefs in order to get rid of dangerous individuals. However, after more than a century we cannot know whether Manuelito and Ganado Mucho believed their accusations. We do know that in cultures possessing witch-beliefs people recognize that one person's interpretation of another's actions may be so coloured by feelings of malice that he may come unjustly to believe the other is a witch. Where witchcraft accusations are related to political competition, there is evidence that people believe persons may deliberately accuse one another falsely. And there is evidence from some indigenous societies that such false accusations did take place.

---

5. Paul Bohannan calls them 'extra-processual' or 'extra-constitutional' events, and likens them to 'witch-hunts' and other activities in the USA, such as the McCarthy hearings (Bohannan, P. 1958: 1, 11; Cardozo 1970).

Case VIII-3

A North American example in which some members of the community came to believe that witchcraft accusations were being made falsely, in order to discredit rivals, concerns Handsome Lake, the prophet and founder of the Longhouse Religion of the Iroquois. In colonial America the Iroquois were a powerful confederation of six tribes living in Pennsylvania and New York State (Jennings 1984), but by the close of the American Revolution their power had been destroyed and they had suffered a severe decline in numbers and lost most of their lands, and occupied a few small reservations in the new USA and in Canada. On their reservations they experienced a drastic decline in morale and in material conditions. Their communities became characterized by drunkenness, violence, and household instability, and accusations of witchcraft increased.

Beginning in 1799, Handsome Lake, one of the *sachems* (chiefs) of one of these reservation populations, the Allegany band of the Seneca tribe of the Iroquois, had a series of visions. These set out a new moral code for the Iroquois and a pattern of accommodation with White society, and revitalized traditional Iroquois rituals. Handsome Lake's early activities as a prophet were concerned particularly with seeking out witches, whom he held responsible for creating social disorganization, drunkenness, fighting, and marital discord. He identified certain individuals[6] as witches, and some of these who did not confess were killed. Among those he accused was his nephew, Red Jacket, an influential Seneca *sachem* who held the office of Speaker of the Seneca Nation. On public trial before the general council of the Seneca, Red Jacket robustly defended himself and accused Handsome Lake of manufacturing charges of witchcraft for political purposes. By a small majority the council judged him not guilty, but it endorsed Handsome Lake's stand on witchcraft in general. (Handsome Lake sought revenge by announcing Red Jacket's Sisyphean punishment, revealed to him in another vision. After death, he would be condemned to fill a wheelbarrow with a heavy load, push it to a certain place, and unload it, for all eternity.)

For a while Handsome Lake became the most powerful person in the Seneca Nation. He continued to hunt out witches. Some old women were killed, and others confessed and were given absolution. Gradually popular sentiment turned against him. He was accused of using witchcraft allegations to remove political rivals, and of becoming a dictator. He alienated some of his most powerful kinsmen and supporters, some of

---

6. 'Sundry old women & men of the Delaware tribe, and some few among (his) own nation.' (Quoted in Wallace 1972: 259.)

A Deed without a Name

whom declared that they no longer believed in witches. The chiefs of the Allegany band officially resolved no more persons should be executed for suspected witchcraft, and Handsome Lake lost his paramount political position; but he remained a religious leader, and the religion he founded helped the Iroquois accommodate to their new situation (Wallace 1972: 254–62, 285–96; Wallace 1978).

Accounts of Handsome Lake consider him sincere in his witchcraft beliefs and accusations. However, some members of the community believed he was using false witchcraft allegations as a political weapon. The community recognized that witchcraft accusations could be fabricated in order to achieve or retain power.

In the career of Handsome Lake again we see witches being accused not only of causing personal misfortunes, but of causing communal ills. This occurs during a period of drastic social change, and the association of communal misfortune with witchcraft is formulated by a new authority, the prophet, who is a product of the changing social situation. Association by prophets of the witch with public evils may be a common consequence of social disruption in small-scale societies. Contemporaneous with Handsome Lake, among the Indians of the Ohio valley, the Shawnee prophet Tenskwatawa ('The Open Door'), brother of the great Indian leader Tecumseh, was instigating witch-hunts against Christian Indians, claiming their witchcraft was responsible for the breakdown of community life (Edmunds 1985: 5, 24–5, 39, 42–7, 85, 97)).

Some scholars have argued that witch-hunts do not take place in tribes and indigenous states, because they lack both a concept analogous to that of the Christian's Devil, and powerful classes and specialist religious institutions that can manipulate it for political ends (Larner 1985: 88, 90, 128, 130). Witch-hunts occur in more complex, peasant-based societies, such as those of Late Medieval and Early Modern Europe; and analogous activities, such as political 'witch-hunts', may occur in modern industrial societies. However, if we mean by a witch-hunt a highly public procedure of searching out persons held responsible for causing, in some mystical way, social ills, some of whom may have been unsuspected, then we find this in Handsome Lake's and Tenskwatawa's searching out of witches, and possibly in the actions of Manuelito and Ganado Mucho.

All three examples concern situations involving unwelcome social change and disruption due to European colonization. When indigenous societies are subjected to such conditions, a common result is an increase in witch-finding, often associated with prophets, and this may take the form of new, organized anti-witch movements. (We observed this development in Chapter 4 in the case of the Lele, and it will be discussed in some detail in the concluding chapter.)

However, it is possible that widespread misfortune, such as famine or epidemics, could promote increased anti-witch activity even in precolonial times. The existence of the Nupe *ndakó gboyá* and other witch-finding associations supports this contention. Evidence from African Kingdoms such as the Azande, Zulu and Ndebele suggests witch-hunts sometimes took place as the result of perceived attacks on legitimate political authority. The Ndebele case suggests they might be deliberately fostered by kings in order to eliminate possible political competitors. However, the Tiv data show that it is not necessary to restrict our comments to states with powerful monarchies. If Tiv witch-hunts occurred in precolonial times, they were associated with the pursuit of power and wealth. The Efik example that concludes this chapter demonstrates the operation of public witch-hunting within a non-state society, albeit a very complex one, in a context of intense competition to acquire wealth and power.

Case VIII-4

A famous precolonial witch-hunt is that attributed by Bryant to the great Zulu king, Shaka. Shaka wished to test the veracity of his witch-finders, so he secretly sprinkled blood around his kraal and then called upon them to expose the witches responsible for this sacrilege. All accused innocent persons, except for Songquoza, son of Ntsentse, who identified Shaka as responsible. Observing Shaka's approval, Nqiwane of the Dlamini clan seconded Songquoza's diagnosis. Shaka rewarded these two, and had the others executed. Although a fraud, this case illustrates a royal procedure for searching out witches. It is possibly the most famous South African witch-hunt, because Rider Haggard incorporated Bryant's account into his romantic novel of the Zulu kingdom, *Nada the Lily* (Haggard 1963: 68–78).

Bryant also records that on one occasion Shaka summoned some three or four hundred witch suspects, mostly women, to his capital and had them all killed (Bryant 1929: 650–1).

These witch-hunts differed from those of European history in being more restricted, involving fewer victims. This is because they lacked the specific ideological dimension provided by the Christian European conception of the witch, that required suspects to name those who were their confederates in a universal conspiracy; and because these indigenous societies lacked the social differentiation and the political and economic complexity of Late Medieval and Early Modern Europe, where specialist institutions staffed by professionals sought out witches. These points will be amplified in the following chapters.

Accusing Political Competitors – the Efik of Old Calabar

Our final example is from the Efik of Old Calabar (Forde 1956; Jones 1956; Simmons 1956). It illustrates many of the themes developed in this chapter, and also provides some unique documentation on the political use of witchcraft accusations in an indigenous society where there was competition for positions of great power and wealth. The Efik are an Ibibio people occupying the estuary of the Cross River in southeastern Nigeria. They have a strong belief in witches. Efik witches can be of either sex. They cause sickness, death, and loss of wealth, and form a society that meets at night to carry out these activities. Its members leave their skins behind to deceive relatives into believing they are asleep. Membership of the witches' society is obtained by sacrificing a relative, or by giving his wealth to the society. In the past, suspects were made to submit to a poison ordeal by drinking eight ground Calabar beans (*Physostigma venenosum*) mixed with water. Innocent suspects would vomit, and if they then continued to be affected by the poison they were given an obnoxious concoction to make them vomit further. Guilty persons were killed by the ordeal, and their bodies were thrown into the forest after their eyes had been removed (Simmons 1956: 21–2).

In the seventeenth and eighteenth centuries, owing to their strategic location at the mouth of a river in the Bight of Biafra, the Efik became middlemen in the slave trade, and in the nineteenth century in the palm oil trade that replaced it. They had a number of closely interacting settlements called 'towns', the most important being Duke Town and Creek Town, which collectively were known as Old Calabar. As a result of the European trade, Efik communities were large (in the mid-nineteenth century Duke Town had a population of 4,000 and Creek Town of 3,000), and important traders were very rich and powerful. Trade was operated through credit with Europeans, and important men received extensive credit from European traders (Simmons 1956: 5).

Efik towns contained a number of trading 'houses'. Each consisted of a core of patrilineally related free men, with their dependants and slaves and the descendants of previous generations of such persons. Each town had a head, to whom Europeans gave the title of 'King'; but, the wealth and power of important men notwithstanding, Old Calabar did not have a state form of organization. There was no common set of hierarchical, bureaucratic institutions. Government within the local community and between communities was organized through a secret society[7] known as

---

7. 'Secret' in the sense that its members possessed secret knowledge and carried out secret activities, not in the sense that membership was a secret. Anthropologists commonly prefer the term 'cult association' or something similar, in order to make this point.

Egbo or Ekpe (Jones 1956: 135–48; Simmons 1956: 16–18). Membership within Ekpe was graded, with the most senior and important men of Old Calabar achieving the higher grades and using the society to maintain peace between trading houses, regulate trade, and make and enforce community laws.

Intensive competition for power and wealth generated conflict between important members of a trading house. This was expressed in the accusations that followed the death of an important member. Competitors for trade and power accused rivals of causing his death by witchcraft and sought to have them submit to the poison ordeal, whilst trying to avoid being placed in such a position themselves. Tensions were particularly intense on the death of the head of a powerful house, when there would be competition to succeed him, and most acute when the deceased was also 'King' of the community. Witchcraft accusations and counter-accusations involving powerful individuals were an important aspect of Efik politics. However, it must be remembered that they were taking place between persons with a developed and genuine belief in witches, and consequently were not necessarily merely deliberate manipulations of the political process in order to discredit or destroy a rival. Often the accuser genuinely must have believed his accusation.[8] But the direction of accusations was influenced strongly by the structure of Efik society and the pattern of intense competition within it (Jones 1956: 150–4).

Because of the importance of the European trade to Efik society, a pidgin English developed as a trade language (Forde 1956: vii). Because of the need to record transactions, leading Efik became literate in this language, and not only kept records, but sometimes diaries and journals. One of these diaries, that of Antera Duke, an important member of the Duke house of Duke Town and a leading Efik trader and a senior member of Ekpe, has been preserved, although incomplete, and provides a fascinating account of Efik society in the late eighteenth century as perceived and lived by one of its most important members (Forde 1956: 27–65). I have taken a selection of entries relating to witchcraft accusations to give a unique insight into the operation of witchcraft and politics in eighteenth-century Old Calabar as given by the matter-of-fact recording of a man who was involved in them.

We start in July 1786, when the head of the Duke house, who is also 'King' of Duke Town, is ill. (The entries are from the English translation of the original pidgin.)[9]

---

8. This is a point made frequently by Mair concerning societies with witchcraft (Mair 1969).
9. The pidgin version is given in Forde 1956: 79–115.

Case VIII-5

2.7.1786
About 5 a.m. at Aqua Landing, and a fine morning. I went to see Duke who was a little sick. At 8 o'clock at night we all took two goats to go and 'make doctor' with Duke.

3.7.1786
About 5 a.m. at Aqua Landing and a fine morning. I went to see Duke, who is sick. After 1 o'clock we all went to Duke's yard to eat the goats we used to make doctor and at 7 o'clock at night Duke was very bad.

4.7.1786
About 4 o'clock in the morning Duke Ephraim died. Soon after we came up to look where to put him in the ground. [When the 'King' of a community died he was always buried in secret (Simmons 1956: 25).]

5.7.1786
About 5 o'clock we put Duke in the ground. Nine men and women went with him, and we all looked very poor. Captain Savage arrived. [Nine men and nine women slaves have been killed and buried with Duke.]

10.7.1786
... All the Cobham Town gentlemen met at George Cobham's yard with our family to chop doctor with us, and return. At 9 o'clock we walked up to Long Dick's cabin with our gentlemen to drink our family doctor. [In the last entry, those persons involved have sworn an oath that they did not kill Duke, by witchcraft or magic, by taking the poison ordeal (Forde 1956: 75). This continues, with members of other 'houses' and other communities.]

11.7.1786
... at 9 o'clock at night we chop doctor with the Henshaw Family at Long Dick's cabin.

18.7.1786
... at 3 o'clock after noon King Aqua came down with 157 men and 16 women and girls; we told him to walk up to George Cobham's, then we and all the Calabar gentlemen went to meet here and drink doctor with Aqua.

19.7.1786
... our family met with ... King Aqua to drink doctor with him and at

Witchcraft, Power and Wealth

7 o'clock at night our family met Willy Tom Robin to drink doctor with him. [The death of an important Efik was not formally announced for a period of from six months to one year (Simmons 1956: 24–5). Duke's death is announced in November, and the mourning observances begin.]

4.11.1786
I went on board Captain Fairweather and saw Efi(om) come up with 3 canoes and play Ekpe in them. After they came up to the landing with Ekpe all the men and women were crying in the town for Duke. I sent my people for wood in two canoes.

8.12.1786
... I went down to the landing to get all the guns ready and we fired 28 great guns ashore, one for each ship. We had shaved our heads first, and we wore fine hats and fine clothes and handkerchiefs. All the Captains and we gentlemen had dinner at Esim's house. [Shaving the head means that the mourning for Duke is ended (Forde 1956: 76).]

24.3.1787
... at 12 o'clock Willy Honesty called all the gentlemen to meet in Egbo Cobham's cabin to decide who we will make King of Old Calabar, and after 7 o'clock at night all of us town gentlemen met at Coffee's cabin to settle every bad quarrel we had had since our father died. So we killed 2 goats. [Here, Duke's successor as head of Duke Town is about to be selected (Forde 1956: 77).]

15.6.1787
... John Cooper dashed Esim and Egbo Young 2 great guns, and we came ashore. We hear that King Aqua has made all his wives drink doctor. So 11 wives died from drinking doctor. John Cooper dropped down river about 8 o'clock at night. [King Aqua has made his wives undergo the poison ordeal. We are not told if this is related to the death of Duke.][10]

---

10. Dr Rosemary Harris has suggested to me that the strength of the Calabar bean concoction might have been varied according to who was taking it. None of the important men who took the oath concerning Duke's death appear to have died, whereas in this case eleven wives probably represent a mortality of at least 20 per cent (personal communication). In contrast to the situation with Coffee Duke on 22.10.1787, the previous situations involving important men are more formal and are not openly situations of competition, and emphasis appears to be on swearing an oath rather than on taking an ordeal.

A Deed without a Name

26.9.1787

About 6 a.m. at Aqua Landing, with a fine morning. This day there were 9 ships in the river. Willy Honesty and all of us went on board the . . . British ship to settle comey palaver . . . Old Otto Duke's daughter drank doctor with her husband's wife. [In this entry there has been an accusation of witchcraft between co-wives.]

22.10.1787

. . . we heard that all the Creek Town gentlemen had come with Willy Honesty to King Ambo's place; so we and all the men and women went to King Ambo's palaver house. Then Willy sent on board Captain Fairweather to call Coffee Duke, and he came ashore with Captain Fairweather. So the Calabar (people) began to ask him what made him run away on board. After these words he says he heard that we four had drunk doctor to kill him. So we told the Calabar that we had drunk doctor (and sworn) that not one of us would go to his house because we heard Duke's sister say that he had killed Duke; (we said) let him drink doctor with Duke's sister before we settle; at the same time we sent to call Duke's sister, and she came to the palaver house. Afterwards she began to bring the dispute to all the Calabar . . . and she says she wants to drink doctor with Coffee. So everybody asks Coffee to drink doctor with her, but Coffee would not drink doctor. So we tell Willy Honesty to send two Ekpe drums to carry Coffee to his house, and we say we four will not settle with Coffee if he does not drink doctor. [Some important men, including Antera Duke, have tried to make a prominent member of the Duke house, Coffee Duke, submit to the poison ordeal, because he is accused of killing Duke by witchcraft. The accusation is made by Duke's sister, and the fact that she is willing to undergo the ordeal with Coffee demonstrates her belief in its veracity.

Unfortunately we do not know whether Coffee Duke escaped the ordeal or was forced to submit. If so, he survived. Nearly three months later we have the following entry.]

8.1.1788

. . . One of Willy Honesty's men came and told me that Coffee Duke has said he will come here and set fire to all his houses. So I took 2 Ekpe drums and blew to forbid any men to sleep in the houses. At the same time I heard that Ndem Duke's son had died from a bad sickness. [A 'bad sickness' is one indicating witchcraft or violation of an oath or the operation of bad medicine (Forde 1956: 78).

The diary ends at the end of this month (January 1788). In the meantime other prominent men have died, and the process is repeating itself.]

3.1.1788
... I went on board to get some brandy for two of my brothers, and at 3 o'clock after noon my two brothers and my Ephraim Abasi went away from Orroup. At the same time we heard that Captain John King Ekpe[11] had died in the Old Town palaver house in front of all the gentlemen.

7.1.1788
... Robin John sent one of his old men to come and say that everybody is wanted to go and take doctor to see if he killed Captain John Ekpe or not. [All the important men have been required to swear an oath that they did not kill Captain John Ekpe . . .]

When witchcraft is associated with positions of marked power and wealth behaviour towards an accused tends to undergo a change. Instead of being acquired gradually by a process of stigmatization and labelling within the community, the status of witch can be attained much more rapidly by being imposed from above, from a position of power. Witchcraft is often regarded as a serious offence, striking against the heart of the community or at large numbers of its members rather than at a few individuals. And the witch-image may be imposed on some accused, instead of being acquired over a long period of time.

However, in societies where power is not diffused widely throughout the community, but is concentrated in the hands of a few officials or leaders, the majority of 'witches' are still persons who have come gradually to have the witch-label fixed to them by the community, and are members of those categories that traditionally are most at risk. This is because leaders and officials remain members of the community and hold its beliefs and are influenced by its opinions. For the same reasons, new leaders, such as prophets, arising under conditions of social disruption and personal anxiety, are likely to blame as witches persons already suspected by the community. But here also, because of the role of the prophet, labelling can be shortened. But the witch's offence is not *being* a witch. It remains the offence of *using* mystical power for antisocial purposes.

When we talk of the political use of witchcraft accusations, we do not mean only their use to destroy political competitors. We mean all the ways in which witchcraft beliefs and accusations may be used within the context of groups or individuals, or the representatives of institutions, seeking to attain or maintain power within the public (as distinct from the private or domestic) sphere of social behaviour (Swartz, Turner and Tuden 1966: 4–8). In the next two chapters I shall examine the witch-idea in Western

---

11. Captain John King Ekpe is probably the head of Ekpe in Old Town.

Europe from Classical times to the end of the 'Great Witch-Craze' of the sixteenth and seventeenth centuries. A fruitful way of examining the development of the European conception of the diabolical witch – the witch that is in league with the Devil – in the Late Middle Ages, and the subsequent witch-hunts of Early Modern Europe, is to examine the role of power and competition for power in these developments.

The diabolic witch was a consequence of powerful individuals and institutions, and the representatives of these institutions, competing for power and seeking to maintain and increase their power. This involved competition between institutions such as Papacy and Monarchy, between individuals such as individual popes and kings, and between individuals and organizations within the same institution, such as different religious orders within the same Church, as well as between competing social groups, such as different Churches or religious organizations. Through the process of political competition witchcraft beliefs became incorporated into the mainstream of Christian theology, as influential intellectual élites linked them with the worship of the Devil.[12] As happens when witchcraft accusations are associated with power and wealth in societies less complex than those of Late Medieval Europe, the process of labelling suspects became shortened, the witch-image was imposed on the accused, witchcraft became a public offence, and the emphasis came to be upon punishment and not upon reconciling 'witch' and accuser.

It is this common pattern of developments in situations involving powerful individuals and institutions that entitles us to treat the witch of tribal societies and the diabolical witch of Late Medieval and Early Modern Europe as aspects of the same phenomenon, although the developments are carried to a greater complexity in Europe, where *to be* a witch became an offence. It entitles us to compare the witch of traditional local communities with that created by Europe's power-holding classes.

---

12. This is not the only interpretation that is made of the creation of the diabolical witch. (See for example Cohn 1976: ix–xiv.)

—9—

European Witchcraft: *Maleficium* and Demonology

Classical Europe, the Dark Ages, and the Early Middle Ages

We have seen that the witch-belief is found in many societies. It was present in the Classical societies of Mediterranean Europe, which had such an important influence on later European cultures. In Classical Greece and Rome there was the belief that certain women could change themselves into animals and leave their sleeping bodies in order to fly about at night. They pursued their own selfish interests, and they used spells and potions to cause illness in people and animals. They held nocturnal meetings, and were associated with the night, the moon, and deities such as Diana, goddess of the moon and of hunting, and Hecate, queen of the spirits of the dead. The classic image of the witch in Graeco-Roman mythology is of a hideous, dishevelled woman in rotting shrouds attending nocturnal meetings to invoke the spirits of the dead (Baroja 1965: 26–40; Russell 1981: 29–32).

In Roman folk belief some women could transform themselves into owls in order to consume the entrails of their living victims. In *The Golden Ass*, the satire from the second century AD by the Roman writer and philosopher Apuleius, the hero Lucius spies on the witch Pamphilë as she changes into an owl to fly to the house of a lover. She smears herself with an ointment and mutters a charm. Gradually her limbs become covered in feathers, her arms become wings, and she flies away hooting over the roof-tops (Apuleius 1950: 88–93; Baroja 1965: 36–9). Belief in such *striges* may have been common among the ordinary people, but the Roman governing classes were sceptical, and they were not recognized in law. On the other hand malevolent use of magic was recognized, and there were frequent prosecutions for this offence (Baroja 1965: 18–20; Cohn 1976: 207–8).

Similar beliefs existed among the pagan Germanic and Slavonic peoples of central and northern Europe (Baroja 1965: 47–57). The Germans believed in women called *striga* or *stria* who used their occult powers to indulge in cannibalism. Like the witches of Shakespeare's

Macbeth, they met around cauldrons (Baroja 1965: 45, 59–60; Cohn 1976: 708). Early German law set a fine to be paid by any *stria* who was proved to have devoured a man, and another to be paid by any person who accused a free woman of being a *stria* and was unable to substantiate the accusation (Baroja 1965: 59–60; Cohn 1976: 208). The ritual position of women in German society was ambiguous. They were venerated and respected, but they were feared because it was believed that they could use powerful spells and might possess special occult powers (Baroja 1965: 48). The Germans vilified their enemies as witches or the descendants of witches (Baroja 1965: 49–50). In pagan Scandinavia occult power was associated with women, and particular families became associated with witchcraft (Baroja 1965: 47). In ancient Gaul witches were believed to be numerous (Baroja 1965: 52). Among the Celts and Teutons witches could change their form, and they used their magic to influence the weather, cause or prevent sickness, and direct love (Russell 1972: 54–5).

In western Europe, witchcraft was included in the general category of *maleficium*, any kind of crime or evil-doing, and someone using magic for malevolent purposes was treated like any other criminal. However, from the fourth century *maleficium* was used in official documents to mean harm caused by occult means. It covered various kinds of occult practices, and this usage persisted throughout the Middle Ages (Cohn 1976: 142; Russell 1972: 13; Russell 1981: 52–3). During the Dark Ages *maleficium* was a personal wrong, an offence against an individual and his kin, and invited retaliation by the offended party. Where the community or the political group was more strongly organized, attempts were made to enforce settlement of disputes involving *maleficium* through payment of compensation. Thus witchcraft was treated generally as a private offence between the parties involved, to be settled according to socially approved procedures. In some circumstances death was considered the proper punishment for the offence – for example, for death caused by *maleficium* when the offender was a slave or a serf; when the witch admitted guilt; or when compensation had not been paid (Cohn 1976: 149).

Case IX-1

Gregory of Tours records a sixth-century example of witchcraft from the Franks that involved Queen Fredegond, wife of the Merovingian King Chilperic. In 580, two of Fredegond's sons died and she accused her stepson, Chlodovic, of employing the mother of his mistress to kill them by witchcraft. The woman was tortured until she confessed to the murders. This persuaded Chilperic of Chlodovic's guilt, and he abandoned him to the vengeance of Fredegond, who had him assassinated. Meanwhile the accused witch had withdrawn her confession. She was burned alive. Her daughter also was killed.

When another of Fredegond's sons died, in 583, she accused the prefect Mummulus, a personal enemy, of employing some women of Paris to kill him by *maleficium*. The women were tortured until they confessed to the murder of the prince, and of many other persons. They were executed. Some were burned, others broken on the wheel. Although Mummulus was tortured he confessed only to using magical salves and potions to try to obtain the favour of the king and queen. He was exiled, and died as a consequence of his torture.[1]

This example is not typical of the general run of cases of *maleficium* of its day. It concerns a monarch (and a particularly ruthless and politically astute one at that). We have seen that witchcraft against kings or other powerful persons is treated as a more serious offence than it is between ordinary folk. Use of torture against a suspected witch is atypical of this time, and is a reflection of Fredegond's status and personality. However, killing in retaliation for witchcraft was not rare, and was an aspect of private vengeance (Baroja 1965: 53; Cohn 1976: 49). It appears that King Chilperic withdrew his support from his son, thus giving up any claim to vengeance for his death and enabling Fredegond to have him murdered with impunity. (In effect, he renounced the relationship between himself and Chlodovic.) Once again we see witchcraft accusations used politically to attack rivals and political enemies. It appears that Fredegond believed her accusations, but their direction was influenced by her personal relationships, and, as she was queen, political factors played a particularly important role in determining these relationships.

Malefici (witches) could cause sickness or death in humans and animals. They could influence a person's emotions to make him or her fall in love. They could cause impotence.[2] You employed them to improve your position at the expense of your neighbours. A particularly dreaded form of *maleficium* was storm-raising, directed at a person's crops. (As we have seen, witchcraft beliefs are especially used to explain misfortunes that uniquely affect one person and not another (Baroja 1965: 95–6; Cohn 1976: 150–4; Robbins 1959: 330–7)).

Maleficium, in this restricted, occult sense, perhaps can best be defined

---

1. This example is compiled from Baroja 1965: 53; Cohn 1976: 148–9. There are differences between the two accounts. For example, Baroja says that the second incident occurred in 578, and the first incident was 'earlier'.

2. See for example the account of witchcraft and impotence in Huxley 1971: 127–8. Although taken from the *Malleus Maleficarum*, it applies to this period also.

A Deed without a Name

as bad magic[3] or sorcery (Russell 1981: 18–36, 42–54, 63–71). However, when suspicions and accusations of *maleficia* built up against an individual who was considered to act out of envy and malevolence, that person came to be regarded as evil by the community, as were Dichtlin and Anna in Case VII-1 and Stürmlin in Case II-1 (although both are from a much later date). She had become a witch.

Christian clergy equated magical practices with paganism, which they identified with the worship of demons. As Christianity spread and became politically powerful, official church attitudes influenced secular authority. In 743, the Frankish king Chilperic III condemned pagan and magical practices. Charlemagne issued edicts condemning all kinds of witchcraft as arts of the Devil. Practitioners and their clients were to be treated in the same manner as murderers, poisoners and thieves. In 805 he authorized limited use of torture against sorcerers. In 873, Charles the Bald declared that makers of *maleficia* and poisons and their clients, male and female, should be exterminated. However, these laws could not be systematically enforced (Baroja 1965: 41–6, 95–6; Cohn 1976: 157–9; Russell 1972: 66–73; Russell 1981: 39–40).

With the break-up of the Carolingian Empire, Church-State relationships ceased to be as close. Although ecclesiastical authorities increasingly defined *maleficium* as a form of heresy, equating it with the Devil, secular authorities continued to regard it primarily as a secular offence (Russell 1972: 70–2). For four centuries after the Carolingian period, until the fourteenth century, witchcraft was not treated as heresy. Although the church was opposed to *maleficium*, it treated certain forms of witchcraft as superstition – as an illusion believed by the ignorant. As part of its campaign to discredit the vestiges of paganism, it derided the idea that witches possessed real occult powers. The *Canon Episcopi*[4] condemned beliefs in women who rode out at night on the backs of beasts in the company of the pagan goddess Diana, or transformed themselves into animals. Such beliefs were dangerous illusions, created by the Devil. The *Corrector* of Burchard, archbishop of Wurms, written at the beginning of the eleventh century, was even more sceptical about the supposed activities of witches and denied the existence of women who passed through closed doors at night to kill and cannibalize Christians, flew through the air, or transformed men into animals (Baroja 1965: 61–3; Briggs 1989: 17 and FN 25; Cohn 1976: 151–3; Russell 1972: 77–82; Russell 1981: 54).

---

3. See Baroja 1965: 23–4 on black and white magic.
4. The *Canon Episcopi* was written about 906, but was held to be a canon of the Council of Ancyra in 314. Consequently, in the Middle Ages it had a great influence on ecclesiastical thought regarding witchcraft (Baroja 1965: 60, 62; Robbins 1959: 74–7; Russell 1981: 53–4).

European Witchcraft: Maleficium *and Demonology*

There are few accounts of trials for *maleficium* prior to the fourteenth century, but documents such as the *Canon Episcopi* and the *Corrector* suggest a widespread belief in witches, especially among the common people. In Chapter 7, I discussed factors that restrict accusations of witchcraft. The same conditions operated in Europe in the Dark Ages and in Early Medieval times (Cohn 1976: 160–3). Also, under the legal systems then operating, accusations of *maleficium* had to be brought privately. Accusers had to amass proof of the offence and conduct the case themselves. The very nature of the supposed offence makes witchcraft particularly difficult to prove, and the structure of the local community and the different interpretations its members placed upon an event meant that many accused were likely to have defenders as well as detractors. Consequently only persons widely labelled in the community as witches were likely to be accused before the law.

In the absence of a confession or of clear proof the accused might be subjected to an ordeal, such as holding a red-hot iron, or keeping an arm in boiling water for a specified time.[5] When the bandage was removed a few days later, if the injury showed no scar they were declared innocent (Cohn 1976; 161–2). Should the ordeal demonstrate the accused's innocence, or the judge remain unconvinced of their guilt, under the law of the *talion* the accusers were likely to suffer a penalty equivalent to that exacted from the accused had they been found guilty. Under these circumstances, appeals to law in witchcraft cases would have been rare. This does not mean that accusations were not made within the community, and there is some evidence from Early Medieval times of popular action against suspects (Cohn 1976: 239–43; Larner 1981: 193–5; Sebald 1978: 65).[6] As judicial procedures changed, we find more depositions formally being made against *malefici*.

The 'Classic' European Witch

In Europe during the medieval period, in contrast to the status of witchcraft in other societies, the witch gradually became incorporated into the mainstream of religion. The idea of the witch became associated with the Christian Devil. She was transformed into the servant of Satan, who gave him her soul and her allegiance in return for the power to commit *maleficium* and the promise of material reward (Cohn 1976: 99–100; Thomas 1973: 521).

---

5. Ordeals for accused members of the aristocracy usually took the form of trial by combat. For ordeals in England, see Thomas 1973: 259–62. Ordeals were abolished in England in 1219, except for the water test for witches (Peel and Southern 1972: 51).
6. See Briggs 1989: 22–7, on informal techniques for dealing with witchcraft.

On much of the continent of Europe, the concern of secular and ecclesiastical authorities came to focus on this aspect of the witch-image. By giving her allegiance to the Devil, the witch was denying her duty to God. Witchcraft came to be treated as a heresy, and convicted witches were burned. In contrast to the situation in England, *maleficium* became of secondary importance in charges of witchcraft. Devil-worship was the main accusation. From the fourteenth century we have the steadily increasing persecution of witches as Devil-worshippers. By the end of the fifteenth century, an elaborate set of beliefs had incorporated witchcraft into Christian theology (Cohn 1976: 99–102; Ginzburg 1990; Ginzburg 1992: 1; Robbins 1959: 369–79, 414–24; Russell 1972: 19, 29, 244–52).[7] They focused upon the witch's pact with the Devil and her attendance at regular witches' meetings ('synagogues' or 'sabbats') at which the Devil was worshipped.[8]

The stereotypical witch was a woman. The Devil appeared to her in human form, or sometimes as an animal, usually at a time of personal distress. He promised material well-being and occult power in return for her allegiance. The wealth rarely materialized – Satan was the great deceiver; but the witch received the power to commit *maleficium*. She formally renounced God, Christ, and the Christian Faith, and pledged herself to serve Satan, who put his mark upon her to show that she was his, a parody of the invisible mark of baptism that made a person into God's own.

With her powers the witch could cause illness and death in people and animals, cause sterility and impotence, and interfere with the weather to her neighbours' detriment. It was believed that she found this an adequate reward, as she was wholly malignant (Cohn 1976: 99–100: Russell 1972: 253).

Witches met at sabbats in order to worship the Devil and plan their evil activities. These were nocturnal affairs that ended at midnight or at cockcrow. Ordinary sabbats were held on Fridays. They were attended by all the witches of a locality, and took place in churchyards, at crossroads, or at the foot of a gallows. Four times a year, special ecumenical sabbats were held, attended by all witches. (At the witch-trials of Dauphiné, held between 1427 and 1447, they were said to be held on Holy Thursday, Ascension, Corpus Christi, and the Thursday nearest Christmas (Russell 1972: 247)). They were often held on the summits of

---

7. See Lewis 1976: 96–8, for a discussion of 'main' and 'marginal' cults. In his terms, Continental witchcraft had ceased to be a 'marginal' cult.
8. See Larner 1985: 79–84, where she differentiates between patterns of witchcraft, and designates the Continental pattern described here as 'Sabbath witchcraft'. The English pattern that I have discussed in Chapters 3 and 4 she describes simply as *Maleficium*.

high mountains. Witches anointed themselves and their animate or inanimate steeds with a salve made from the fat of murdered infants to fly to sabbats on animals, spits, shovels, sticks or broomsticks. To deceive her spouse, a witch left behind in their bed a stick that assumed her appearance.

The Devil presided over the sabbat in a form half man and half goat. The witches made homage to him and repeated their renunciation of Christianity, and the Devil delivered a sermon in a blasphemous parody of divine service. The witches adored him by kissing his anus – the notorious *osculum infame*. There followed a meal of revolting substances – a parody of the Christian eucharist. The meeting ended with a sexual orgy that involved incest, sodomy, and intercourse in the manner of animals, while the Devil copulated with everyone in turn. Finally the assembly broke up, and the witches returned home enjoined to commit every possible form of *maleficium* against their Christian neighbours (Cohn 1976: 100–2; Russell 1972: 144–251; Russell 1981: 37; Sebald 1978: 68–9).

On much of the Continent, therefore, theologians and the governing classes came to regard witchcraft as the most extreme type of apostasy. It was a perversion of Christianity, its uttermost antithesis. As in other societies, witches had an organization and activities that inverted or caricatured the normal world. Their meetings were a perversion and a parody of Christian worship. They desecrated the Cross and the sacraments to demonstrate their rejection of Christianity. They murdered babies – their own and those of others, cannibalized their dead bodies and rendered down the residue to make ointments to enable them to fly, to stay silent under torture, and to kill their victims.

Because of the perverted association between witchcraft and Christianity, witches must be former Christians. Non-Christians, such as Jews, although accused of other forms of occult wrongdoing, were not accused of witchcraft. However, the terms 'sabbat' and 'synagogue', both Jewish, demonstrate the medieval view of Judaism as fundamentally anti-Christian (Cohn 1976: 101–2; Russell 1972: 19, 29, 251–2, 266–7). Because the witch was the enemy of God she was also the enemy of humankind. To be a witch was a public offence, and consequently an accusation of witchcraft could no longer be a private matter.

Heresy and Devil-Worship

The development of European witchcraft into Devil-worship was a complex historical process. The European witch-image was particularly a medieval creation, and was formulated largely between the eleventh and the fourteenth centuries. Jeffrey Russell identifies the different elements that contributed to this conception of witchcraft. One of these was 'sorcery' (his term for *maleficia* and related practices). European folklore was another. Ancient elements of folklore were incorporated into the developing image of the diabolical witch, becoming transformed in the process. They included the Teutonic legend of the Wild Hunt. This was the belief in men and women who rode out at night on animals and fence rails in the company of the goddess Diana (who became identified with Hecate, the queen of the night, and Herodias, the evil wife of King Herod (Baroja 1965: 60–6; Cohn 1976: 212–19; Duerr 1987: 12–16, 35–8; Ginzburg 1992: 89–121)). Aspects of pagan religion formed another element. For example, the dates of some important pagan festivals became the dates of ecumenical sabbats (Russell 1981: 50). Other elements were Christian heresy and medieval theology (Russell 1972: 22–6; Russell 1981: 52–71).[9]

Historians, examining the development of the classic European idea of the witch and the related witch persecutions, commonly write of 'witchcraft' as a new phenomenon, a synthesis of traditional beliefs and new ideas, that developed during the Middle Ages.[10] In fact, by our definition, 'witchcraft' existed in European societies long before the fourteenth century. What occurred in medieval times was the transformation of the pattern of witchcraft that we have been examining in previous chapters into a more complex pattern that was part of the mainstream of religious ideas and practices of the time. In order to differentiate this new kind of witchcraft, I shall follow Jeffrey Russell and refer to it as 'diabolical witchcraft' (Russell 1981: 72, 76). Where it is necessary to make the distinction, I shall refer to the pattern of witchcraft described in previous chapters as 'traditional witchcraft'.

The development of the European witch-image appears particularly to have been the consequence of powerful groups', institutions', and individuals' competing with each other, or seeking to consolidate or ensure their power. In other words, it was a *political* development. This conclusion is supported by the fact that the subordinate classes and groups of European society continued to believe largely in the traditional conception of the witch, and were concerned with her as a practitioner of *maleficia* rather than as an agent of the Devil.

The basic features of the European witch-image are: the pact with the Devil; worship of the Devil; sexual orgies; and the sacrifice and cannibalizing of children. These, or similar, ideas have a long pre-

---

9. On the elements that went to make up the demonic witch, see also Ladurie 1987a: 76–7.
10. See for instance Russell 1981: 63–6, who writes of 'sorcery', meaning witchcraft of the type discussed in Chapters 2 to 8, and 'witchcraft', the new Continental belief.

medieval history in Europe, and some are associated with the witch-image in many parts of the world. In the second century, Christians were accused of worshipping the head of a donkey at their secret meetings, killing and eating children, and engaging in promiscuous and incestuous orgies (Cohn 1976: 1–6). Other groups had been similarly accused, such as the Jews in Alexandria in the first century BC and the first century AD (Cohn 1976: 5–6; Russell 1981: 59). Cohn suggests ritual murder and cannibal feasts have frequently been used to stereotype secret organizations or conspiracies perceived as seeking to overthrow established authority (Cohn 1976: 7–8). (However, we should note that these minorities fall within our categories of persons likely to be stereotyped as wicked.) So common are these beliefs, that Cohn argues they may be a projection of repressed human desires and temptations and of unconscious fantasies created in infancy or early childhood (Cohn 1976: 256–63).

The idea of organized Devil-worship appears in 1022, in the trial of a group of heretics at Orleans conducted by Robert II of France. It is significant in being the first medieval execution for heresy, occurring after several centuries in which heresy had ceased to be an important ecclesiastical concern. The accused were literate clergymen who wished to reform the church, some aristocratic laymen, and some women, including nuns. They rejected the Church's claim to mediate between the Christian and God, and argued that salvation was achieved through renouncing the material things of this world, engaging in manual labour, and harming no one (Klaniczay 1990: 73–4). The Church accused them of secret nocturnal orgies in which men and women mated indiscriminately, irrespective of relationship or vocation. They were charged with adoring the Devil, who appeared as an angel of light or a black man or an animal, and denying Christ. It was claimed that children born from their orgies were burned on the eighth day after birth, the day for Christian baptism, and their ashes used in a grotesque parody of the Christian eucharist (Cohn 1976: 20; Russell 1972: 87–8, 90, 93; Russell 1981: 58–9, 62, 65).

Other heretical groups were similarly stigmatized. The Cathars, who were prominent in Europe from the tenth to the fourteenth century, especially in Lombardy, the Rhineland, and southern France, were charged with cannibalism, infanticide, and holding sexual orgies. Walter Map's 1182 account of Catharist ritual, written by an orthodox Catholic, claims the Devil appeared at their meetings in the form of a huge black cat. They worshipped him and paid homage by kissing his private parts (one of the first mentions of the *osculum infame*) (Lea 1957: 200; Robbins 1959: 244–5; Russell 1972: 120–32; Russell 1981: 160–2). Pope Innocent III directed a crusade against the Albigensians, the Catharists of southern France who enjoyed the protection of Duke William IX of Aquitaine, from

1209 until 1229. They survived, which caused Pope Gregory IX to found the Inquisition in 1232, and were not finally suppressed until 1330 (Ladurie 1987b; Robbins 1959: 244–5).

A heresy that became particularly closely associated with witchcraft in the popular mind was that of the Waldensians. The sect was founded in 1173 by Peter Waldo, or Valdes, as a mendicant order that stressed the value of voluntary poverty. Its members became known popularly as 'Insabbatati' or 'Sandalati' ('Sandal-wearers') (Klaniczay 1990: 73–4). In the following decade, when the Waldensians refused to obey the Pope's interdiction against their preaching, they were excommunicated and then condemned as heretics. Taking the Vulgate as their authority they claimed that the Roman Catholic Church was an abomination and opposed its clergy.

They were persecuted, and during the Albigensian crusade many fled to the Piedmont. They were accused of indulging in incestuous orgies at underground meetings presided over by the Devil in corporeal form, and of kissing the hindquarters of a demon in the form of a cat. Some of these claims were incorporated into a bull by Pope John XXII in 1318. This called for the seeking out and prosecution of those who entered into contracts with devils, invoked them in order to commit *maleficium*, and misused the sacraments for sorcery and witchcraft (Lea 1957: 220). From 1365, an intermittent crusade was waged against the Waldensians. In 1487 Pope Innocent VIII appointed Alberto Cattaneo as Inquisitor for the Dauphiné. Cattaneo called for a military invasion of the Waldensians' strongholds. Many were killed. Some were burned as impenitent or relapsed heretics, and others were received into the Church. Under torture, some confessed to taking part in nocturnal orgies. In fact, the moral probity of the Waldensians appears incontrovertible. However, so close had the association been made between witchcraft and the Waldensians, that witches were often referred to as *waudenses* or *vaudois*, and the witches' meeting as a *vauderie*. (Witches were also identified with Catharists, by terms such as *Gazarii*.) (Cohn 1976: 32–42; Robbins 1959: 245; Russell 1972: 219–21; Russell 1981: 75)

Ecclesiastical authorities stressed that Devil-worship was an organized group activity. The basic characteristics of the witches' sabbat – the night meeting with the Devil presiding, the indiscriminate sexual orgy, the murder and cannibalizing of children – were constructed from the prosecution of medieval heresies, and the idea of the sabbat was created on the Alpine borderlands, between 1375 and 1450 (Ladurie 1987a: 72–7). When heretics confessed to these charges it was under torture and in response to the preconceived views of their interrogators. However, cross-cultural examination suggests that some of these beliefs, or similar ones, would be present in popular conceptions of witchcraft. Jeffrey Russell estimates that about 80 per cent of the elements that went to make up the classic European witch-image were already present in folk beliefs. The other 20 per cent were created during the development of the concept of diabolical witchcraft (Russell 1972: 22).

Several writers on witchcraft have claimed that the Cathars hold a special place in the development of European witchcraft (Cohn 1976: 57–8, 126, 128–30, 138). Russell believes their ideology provided a marked stimulus to the notion of Devil-worship. Cathar religious belief was based upon a strongly dualist tradition of Christianity that began with the Gnostics and the Manichaean heresy of the early part of the Christian era (Cohn 1976: 16–17, 57–9; Russell 1972: 123; Russell 1981: 60). The Devil's power was almost equal to that of God. Catharists believed the Devil, the spirit of evil, was creator and lord of the material world. He was the God of the Old Testament. He had trapped the human soul in a material body. Christ was a pure spirit sent by the good, hidden God to teach man how to liberate himself from the material world (Cohn 1976: 57–8; Ladurie 1987b; Russell 1972: 123; Russell 1981: 60). Russell believes the dualism of the Cathars stimulated the latent dualism present in orthodox Christianity, so that the Devil came to play a much larger role in Christian theology.[11] On the other hand, Norman Cohn argues that the increasing medieval preoccupation with the Devil and demons was a consequence of the failure of the Church's early millenarian promise to transform the world. This gave rise to a belief in the ubiquitousness of the Devil, with his demonic minions and human servants (Cohn 1970; Cohn 1976: 60–74). (He also argues that these beliefs are projections of repressed desires created by medieval Christianity.)

Whatever the reason, from the mid-twelfth century medieval theology showed an increasing concern with the role of the Devil. Thomas Aquinas (c.1227–1274) formulated the doctrine of the 'implicit pact', which argued that all heretics are implicitly in league with the Devil, whether or not they conjure him up deliberately – a dictum to be used later with great effect against suspected witches. Scholars also introduced the idea of ritual sexual intercourse between Devil-worshipper and Devil, and since the Devil was perceived as male this was to provide a theological explanation for the predominance of women among those accused of witchcraft (Lea 1957: 200–1; Robbins 1959: 28–9; Russell 1981: 66–70).

From the twelfth century, the more efficient organization of Church and State enabled religious and secular authorities to exercise greater

---

11. See, for example, Russell 1972: 272–89; Russell 1981: 60. I find the argument put forward for the influence of Catharist ideology rather tendentious. The main reason for Russell's stress on the importance of Catharist ideology appears to be his belief that a medieval organization of Devil-worshippers, rebels against society, actually existed.

control over civil and ecclesiastical courts. Because of this, and the introduction of Roman Law into both civil and canon law, the legal system became more severe in dealing with heresy. Harsher penalties were applied, and the active searching out of heretics was promoted. Roman Law allowed the use of torture in the interrogation of suspects, and it was applied increasingly as fear of heresy grew. By the late thirteenth century bishops were encouraged to establish their own inquisitions (formal examinations) for heresy.

The growth of heresy and the increasing centralization of ecclesiastical organization led to the creation of the Papal Inquisition, between 1227 and 1235, to search out heretics. Its powers were strengthened by successive popes. In 1231 the archbishop of Mainz appointed Konrad of Marburg the first Papal Inquisitor in Germany. He was a religious fanatic, and with the assistance of two far less scrupulous individuals, Conrad Torso and Johannes, he produced heretics that were more diabolical than any previously discovered. He found them mainly through denunciations by confessed heretics. Once accused, suspects were tortured to make them confess and denounce others. This procedure produced tales of secret meetings where the Devil appeared as a toad, a goose, a duck, a black cat, a pale thin man with shining black eyes, or a man whose upper half shone like the sun but who was covered in hair from the waist down. There were accounts of the obscene kiss and of sexual orgies. These products of threats, torture, and the inquisitor's leading questions were given papal recognition in 1233 by Gregory IX in a bull entitled *Vox in Rama*, aimed at the Waldensians (Cohn 1976: 23–33; Russell 1972: 159–61; Russell 1981: 69–71).

The Inquisition played a decisive role in the incorporation of *maleficia* and traditional witchcraft into the developing concept of diabolical witchcraft. From the early thirteenth century, inquisitor's manuals often included questions on witchcraft. In mid-century, Alexander IV refused the Inquisition's request for jurisdiction over all cases of sorcery, but allowed it those that involved heresy. Consequently inquisitors introduced heresy into trials for more conventional sorcery by using the theory of the implicit pact and other theological arguments. Although the Inquisition was influential in spreading the idea of diabolical witchcraft, both episcopal and secular courts continued to try witchcraft cases. All authorities agreed that witchcraft must be eradicated, and the increased use of torture in witchcraft investigations, by all courts, was an important influence on the growth of prosecutions. In their turn, trials and convictions helped impress upon the population the developing idea of the diabolic witch (Russell 1972: 156, 204–5, 227–30; Russell 1981: 71).

European Witchcraft: Maleficium *and Demonology*

Political Trials Involving Beliefs that became Incorporated into Late Medieval Witchcraft

The early fourteenth century was marked by a spate of dramatic trials involving accusations of Devil-worship made by powerful persons for political purposes. Between 1300 and 1360 there were ten important politically-motivated trials involving charges later associated with the diabolical witch. (In all of these, the defendants appear largely to have been innocent of the kinds of practices of which they were accused.) Russell suggests that this was the time when the political possibilities of such accusations were just becoming appreciated (Russell 1972: 193–4; Russell 1981: 76). The most notorious trial is that of the Knights Templar, which has been described as 'the scandal of the Middle Ages' (Costain 1973: 159–69). In fact this was only one of three interconnected political trials instigated by King Philip IV of France that involved demon-worship.

Case IX-2

Philip IV ('the Fair') was a powerful, ruthless and ambitious king who appears to have had dreams of becoming the most powerful monarch in Christendom, analogous to the ancient Roman Emperors. France and the England of Edward I were at war, and Philip taxed the clergy for a contribution to the war. This was contrary to canon law, and it brought the king into conflict with the pope, Boniface VIII. Boniface reiterated the papal claim to control over the clergy, but Philip further challenged him by imprisoning the bishop of Pamiers.

In his confrontation with the pope Philip had the support of one of the most powerful Roman families, the Colonnas. Boniface was a member of the rival Caetani family, and his accession to the papacy had made it the most powerful Roman family. He used his position to try to break the power of the Colonnas, so they joined Philip's campaign to discredit him and began a rumour that he was in league with demons. At an assembly of bishops and lords at the Louvre in March 1301, Philip's chief minister, Guillaume de Nogaret, used these rumours to denounce Boniface as a heretic and demand his trial by the Church. The assembly declared him deposed and charged him with being a sorcerer and possessing a familiar spirit.

In September Nogaret and the Colonnas seized the pope at his castle in Agnani. He was rescued, but died within a month of the affair. His successor, Benedict XI, excommunicated Nogaret for his part in the incident. However, Benedict was an old man and soon died, to be succeeded by a Frenchman, Clement V. Clement owed his election to Philip's support, and he made his residence at Avignon.

Philip's ambitions to become the premier ruler of Europe involved merging the three crusading orders of knights – the Hospitallers of St John, the Teutonic Knights, and the Knights Templar – into a single order. The Knights Templar was a powerful and extremely wealthy medieval order that was founded about 1118 to protect pilgrims to the Holy Land and fight the forces of Islam. Over a period of two centuries it had become immensely wealthy and owned more than 10,000 estates and manor houses throughout Europe. The Templars utilized the Order's assets and became the bankers of Europe. The safety of Templar houses enabled them to become sanctuaries for crown jewels, jewellery, and personal and private monies.

The Order lent money to kings, nobles, and great merchants, and was allowed to charge interest on its transactions in the guise of rents, usury being illegal. Taxes were collected and paid through the Temple,[12] which kept a percentage for its services. It used its facilities to transfer funds from one country to another, and issued letters of credit that were honoured everywhere. Templars were employed in diplomatic and political offices, and held influential posts at the courts of prelates and monarchs. In the west the order's headquarters was the Paris Temple, built on land given to it by Louis VII in thanks for its support during the Second Crusade. Of the Temple's 4,000 members in Europe, approximately half were in France.

The Order had remarkable privileges. It paid no taxes on its immense wealth. In 1163 Pope Alexander III had rewarded it for supporting his election by making it an autonomous institution subject only to the pope. Its possessions were under the protection of the Holy See. It could build its own churches and appoint its own confessors. Its privileges and its remarkable wealth and power attracted hostility from both secular and ecclesiastical sources, and the Order itself had become arrogant and ruthless in pursuit of its interests. It maintained strict secrecy about its rites and proceedings, and this helped promote rumours about its practices.

The Grand Master of the Temple, Jaques de Molay, rejected Philip's proposal to merge the crusading orders, and Philip saw the Temple as an obstacle to his plans. Probably more importantly, Philip's realm was almost bankrupt, its finances weakened by costly wars. He determined to seize the Order's wealth; but if he was to do so it would have to be suppressed. To this end, he acted on claims made by a Frenchman, Esquiu de Floyran, that the Templars engaged in Satanic practices. (One version of the story of Esquiu de Floyran says that he claimed he learned of these practices from an apostate Templar, who confessed to him when they were

---

12. 'Temple' was the term for the Templar order.

in prison together.) Philip informed the pope, Clement V, of Esquiu's accusations, and in August 1307 Clement replied that the claims would be investigated. To pre-empt papal action that might prove favourable to the Temple, Philip made secret plans to strike against it, in defiance of the pope's jurisdiction over the Order.

Philip laid his plans with care. His relations with the Temple had always been good, and the Order had no reason to be suspicious of him. It had recently advanced the dowry of his daughter Isabella for her marriage to Edward, heir to the throne of England. Hughes de Pairaud, treasurer of the Paris Temple, was receiver and warden of all the royal revenues. In 1306, the very year before his onslaught on the Order, Philip took refuge in the Paris Temple to escape the anger of the populace of his own capital! (They were enraged at the consequences of his debasement of the currency.)

Philip invited the Grand Master to France to discuss his plan for merging the crusading orders. Unsuspecting, de Molay came prepared to oppose it. On 14 September 1307 the king dispatched secret orders to his seneschals to arrest all the Templars in France on the night of 13 October and seize the Order's archives and property. The instructions detailed the accusations against the Temple. It was claimed that on his initiation into the Order a postulant must formally deny Christ, God, and the Virgin. He had to spit on the Cross three times, and trample or urinate upon it. He had to kiss the mouth, navel and buttocks of the prior and caress him sexually. The Temple was accused of sodomy. It was claimed that the Order adored an idol in the form of a golden calf or the head of a bearded man. In later accusations, this idol was said to be called 'Baphomet', a corruption of Mohammed – an attempt to link the Order with Islam.

Completely unprepared, the Templars were arrested and subjected to inquisitorial procedures involving merciless torture. Initially this was carried out by royal, secular officials. Only after confessions had been obtained were papal inquisitors called in, and they remained subordinate to the royal power. The situation was without precedent in medieval history. Philip felt able to behave in this way because Clement owed his position to him, and because the pope was an ill man. Torture produced confessions to supplement or refine the original charges: Templars adored a demon in the form of a cat, bowing before it and kissing it beneath its tail; they had intercourse with succubi – demons in female form; they made powders from the ashes of their murdered illegitimate babies, and from the ashes of dead Templars, which they administered to novices. The Franciscans and the Dominicans, who hated the Templars, acted as propagandists for Philip, and he stage-managed a number of show confessions, including that of Jaques de Molay. Old and frail, the Grand

Master had broken down under fear of torture. He confessed that the Order had been seduced by Satan.

The accusations against the Templars appear to have been framed particularly by Guillaume de Nogaret, who was a university-trained lawyer. They were cleverly prepared to appeal to all levels of society. They were plausible to the public because they utilized powerful themes from the folk-culture of the time. Three of the accusations had a particular appeal. These were that the Templars worshipped idols, did not consecrate the host when they conducted mass, and indulged in homosexual practices.[13] Similar accusations were commonly made against the Cathars and Waldensians, and de Nogaret may have intended to associate the Templars with these heretics.

The most sensational evidence concerned a talking bearded head made of wood or bone, or metal or silver or gold. It was variously described as painted on a beam, having four legs, having a demonic countenance, being the head of one of the thousand virgins, being painted red, having two faces, being painted on a picture, and wearing a cap. In folk-mythology, such heads were the product of copulation with a woman's corpse. Failure to consecrate the host meant that the mass was ineffective. This implied that the many patrons who had made material benefices to the Templars over two centuries, in return for their spiritual support, had been tricked. The Church preached that homosexuality was the sin that destroyed Sodom. Imputing homosexuality to the Templars implied that the loss of the Holy Land was a consequence of their sexual perversion. The wrath of God would surely fall upon a France guilty of sheltering them and failing to punish their depravity. If they were unpunished, their sacrilege would result in widespread public misfortune.

In 1308 Clement made an attempt to assert his independence. He suspended the inquisitorial powers of bishops and inquisitors, and reserved for himself all decisions concerning the Temple and its possessions. Philip reacted by calling an Estates General in Tours, where his minister, Guillaume de Nogaret, presented the accusations against the Templars as proven facts. The Estates General called for their execution. Philip next convened a great assembly of clergy and laity at Poitiers and threatened the pope with a charge of heresy if he failed to recognize the enormity of the Templars' crimes. To put further pressure on Clement, Philip also publicized the case of Guichard, bishop of Troyes.

Guichard was head of a priory at Provins when he attracted the patronage of Blanche of Artois, the widow of Henry III of Navarre, and her daughter, Joan, who later became the wife of Philip IV. Through their influence he was appointed bishop of Troyes, and became rich and

---

13. Detail on the accusations against the Templars is mainly from Barber 1973.

powerful. Some time later he lost favour with the two women, was persecuted and had some of his property seized. Queen Joan was informed that Guichard employed a Jew to summon a demon to frighten her into dropping the case against him. However, Guichard's innocence was established, and he was reconciled with the Queen. In 1307 Pope Clement V formally recognized his innocence.

In 1308 Guichard's enemies accused him of practising *maleficia* and plotting to poison the king's children and brother. The queen had died in 1305. Philip demanded an enquiry into the bishop's behaviour on the grounds of attempted regicide and commerce with demons. The harassed Clement ordered Guichard's arrest. Royal officials forced charges to be laid before an assembly – the same procedure that had been used against Boniface and the Templars. The old charges concerning Queen Joan were revived and elaborated. Guichard was accused of summoning the Devil to assist him against the queen. It was alleged that the Devil advised him to make a wax image of the queen, baptize it, and mutilate it. She died as a result. Under torture, Guichard's chamberlain testified against him. An indigent fortune-teller, Margueronne de Billevillette, was threatened with torture and promised immunity if she testified against Guichard. She said that she saw him conjure up the Devil with the help of a Dominican friar. Other witnesses testified that Guichard made a compact with the Devil that resulted in the death of the queen.

In 1309 the enquiry was resumed with new accusations and new witnesses. It was claimed that Guichard was the son of an incubus (a male demon) and that he had a private demon that he consulted for advice. He had made a mixture from snakes, scorpions, toads and poisonous spiders to administer to the royal princes. He was accused of multiple homicide. Guichard denied the charges. In April 1311 the commission submitted its report to the pope. Clement had given way to Philip's demands and condemned the Temple, so Philip had lost interest in Guichard and Clement refrained from further action. Guichard's innocence was affirmed by one of his principal accusers, and in 1314 he was freed to become the suffragan bishop of Constance.

Clement capitulated to Philip and set up a commission to investigate the Temple. In France, Templars recanting their confessions were burned as lapsed heretics. However, confessions were only obtained in France and in those parts of Sicily and Italy that were controlled by Philip or Clement, because merciless torture was used only in these areas. In England, for example, any torture that would cause perpetual mutilation or disability of any limb, or create a violent effusion of blood, was not permitted against the Templars.

Philip also required Clement to annul bulls made against him by Boniface VIII, and Nogaret wanted his sentence of excommunication annulled. To further pressure Clement, they proposed that the deceased Pope Boniface should be tried on charges of heresy, apostasy, and crime. In 1310 Nogaret presented Clement with charges against Boniface. Cardinal Peter Colonna alleged that the deceased pope had owned three demons, and a 'spirit' in a ring on his finger. These advised him and helped him in his feud with the Colonna. It was alleged that Boniface had given himself body and soul to these demons. The accusations were substantiated by specially selected witnesses.

At the second hearing of evidence against Boniface, in 1311, three monks charged him with demon-worship. It was claimed that he made a pact with his demons when he was a notary. When he became pope he worshipped an idol containing a diabolic spirit. These were the same kinds of charges that were being levelled at the Temple. As were the Templars, Boniface was charged with apostasy, sodomy, and murder. He was accused of murdering his predecessor and of favouring the Templars and accepting money from them. But by 1311 Clement had yielded to Philip's pressure and suppressed the Temple and absolved Nogaret, and the charges against Boniface were quietly dropped.

On the 2 March 1311 the pope, under pressure from Philip, issued a decree suppressing the Temple. The Order's wealth was given formally to the Hospitallers, but Philip already had obtained much of it. With the exception of those considered to have relapsed, the surviving Templars were distributed throughout various monasteries. On 18 March 1314 four great officers of the Temple appeared on a public platform to be sentenced to imprisonment for life. The Grand Master, Jaques de Molay, and the Preceptor of Normandy, Geoffrai de Charnay, used the occasion to repudiate their confessions and declare the Order innocent of the heresies and actions of which it had been accused. Philip had them burned alive the next day. As he burned, de Molay summoned king and pope to meet their judgement. Monarch and prelate both died within a few months.[14]

Such sensational events helped fix the idea of Devil-worship and the demonological witch in the public consciousness, and increased concern over witchcraft. A fourteenth-century trial that appears to have been influenced by the French political trials is that of Lady Alice Kyteler of Kilkenny, Ireland, in 1324–5.

---

14. Case IX-2 is taken from the following sources: *Boniface VIII* – Barber 1973: 44; Cohn 1976: 180–5; Russell 1972: 187–8. *The Templars* – Barber 1973; Cohn 1976: 75–98; Costain 1973: 159–69; Russell 1972: 194–8. *Guichard of Troyes* – Barber 1973: 44; Cohn 1976: 185–92.

Case IX-3

Lady Alice was a wealthy, much-widowed Anglo-Norman gentlewoman. She had married four times and outlived three husbands, who left their property to her and her eldest son, William Outlaw. (Outlaw was a banker and money-lender, and many of the local nobility were in debt to him.) She was accused by her stepchildren of bewitching her husbands into leaving her their property, and then murdering them. In 1324 they accused her before Richard de Ledrede, bishop of Ossory. Ledrede held a formal inquiry. Some local knights and nobles were involved in the inquiry, but the main witnesses against Lady Alice were her stepchildren. She was accused with twelve other persons, apparently from ruling Anglo-Norman families, and it was claimed that they constituted an organized group of demon-worshippers.

It was alleged that the group denied God, called upon demons, and engaged in all kinds of *maleficia*. Although torture was not allowed in Ireland under English common law, Ledrede, who was a fanatic, had one of Lady Alice's associates, Petronilla of Meath, flogged six times, until she confessed and elaborated on the accusations. She said that Lady Alice had renounced Christ and the Church in order to obtain magical powers. She sacrificed cocks to demons at crossroads, and in particular to her familiar, Robin (or Robert) Artisson. Petronilla and Lady Alice had made love charms using the brains of an unbaptized child and the skull of a decapitated robber. When the thirteen associates held their secret nocturnal conventicles, the candles were extinguished and the elderly Lady Alice and her companions cried out 'Fi, fi, fi, amen!' and fell upon each other in a sexual orgy.

The Kyteler case is one of the earliest involving the idea of a familiar spirit, and the earliest known example in European history of a women acquiring occult power through sexual intercourse with a demon – a feature of later, 'classic' European witchcraft. Robin was Lady Alice's *incubus*, a demon in male form with whom she had intercourse. He appeared as a cat, a shaggy dog, and a black man who was accompanied by two taller comrades, one of whom carried an iron rod. These demons taught Lady Alice how to make magical concoctions.

Bishop Ledrede demanded that the Lord Chancellor, Roger Outlaw, order the imprisonment of the accused. Outlaw was Lady Alice's brother-in-law and William Outlaw's uncle. He did nothing. Ledrede proceeded on his own until the seneschal of Kilkenny, Arnold le Poer, sought to make him desist. Le Poer was a relative of Lady Alice's present husband, who had supported the charges against her. But le Poer supported Lady Alice and accused Ledrede of being a meddlesome Englishman who found heretics under every bed. In this he had the support of some of the Irish bishops.

Ledrede excommunicated Lady Alice, who retaliated by indicting him for defamation of character. Le Poer had him arrested and lodged in Kilkenny jail. His arrest and confinement brought him the support of the local clergy. On his release he had an angry confrontation with le Poer. Lady Alice's allies, led by le Poer, had Ledrede cited to appear before the parliament in Dublin for improperly and unjustly excommunicating her. However the bishop convinced the parliament, and was allowed to proceed against the accused. Le Poer was forced humbly to beg his pardon for wrongly imprisoning him. All the accused were found guilty. On 3 November 1324 Petronilla of Meath was burned alive – the first person in Ireland to suffer death by fire for heresy. Some of the others also were burned, while the rest received less severe punishments. William Outlaw was supported by Roger Outlaw (the Lord Chancellor of Ireland) and Walter Islep (the Treasurer), and on one occasion he answered a summons by Ledrede accompanied by a band of armed men. But finally he was forced to submit and do penance. He was reconciled with the Church after a period in prison. Lady Alice escaped to England with the assistance of her powerful kinsmen. Her stepchildren received their inheritance. Ledrede pursued Arnold le Poer until he had him excommunicated and imprisoned in Dublin Castle, where he died.

It is clear that in this case powerful political and financial interests were at work and influenced support for both sides. Bishop Ledrede was the instrument through whom these interests were fought out. However, he appears to have believed his accusations. He was a fanatic, and as le Poer claimed, he found 'heretics under every bed'. He had visited the papal court at Avignon at a time when the accusations against Boniface VIII and the trials of the Templars and Guichard of Troyes were very recent events. (He was consecrated at Avignon in 1317.) He was familiar with the accusations involved in these cases, and they appear to have influenced his conception of heresy. This was reflected in the framing of the accusations in the Kyteler affair.

Ledrede himself was later accused of heresy by the Archbishop's Court in Dublin, and fled to the papal court in Avignon. When, after nine years, he was able to resume his office, he came into dispute with the archbishop of Dublin. He persuaded Clement VI that the archbishop was protecting heretics. His fanaticism was persistent, and coloured his interpretation of events in which he was involved.[15]

---

15. Cohn 1976: 198–203; Parrinder 1963: 88–9; Robbins 1959: 294–5; Russell 1972: 189–92; Russell 1981: 90–2; Seymour 1992: 25–51. These accounts vary in their description of the relationships within the Kyteler family and of the sequence of events. Where it is necessary to outline the sequence of events I have generally followed Cohn 1976, which appears to use primary sources.

In several of the examples given above, the offence of which the parties were accused was apostasy involving the conjuring of demons. In fact, Boniface VIII appears to be the first person to be charged with raising a demon. In the cases of Alice Kyteler, Guichard of Troyes, and Boniface VIII there were also charges of using the power given by demons in order to carry out *maleficia*; but these were presented as secondary offences. The nature of the accusations was determined by the late medieval Church's attack upon ritual magic as apostasy. Ritual or ceremonial magic was a highly sophisticated, scholastic magic concerned with astrology and divination, believed to involve the conjuring of demons in order to compel their knowledge and assistance. It was a magic of the educated classes. In the thirteenth century, Thomas Aquinas argued that no demon is really coerced. He is in fact seducing the magician into sin. It was argued that in effect the magician enters into a pact, implicit or explicit, with the demon, and in 1320 Pope John XXII issued a bull, *Super illius specula*, which identified the invoking of demons as heresy and authorized the Inquisition to proceed against all sorcerers because they worshipped demons, and made a 'pact with hell' (Lea 1957: 221; Russell 1981: 76). Although *maleficium* had a secondary role in the resulting trials they served to strengthen the idea of an association between *maleficium* and Devil-worship, and hence an association between the village witch and the Devil (Cohn 1976: 164–79; Robbins 1959: 294–5; Russell 1972: 173; Russell 1981: 76).

The Kyteler case introduces the idea of an organized group of heretics meeting together and using their demonic powers to carry out *maleficia*. This belief lies behind the idea of the witches' sabbat. However, the Kyteler case has certain incongruities. It contains in developed form ideas that were important in the later European conception of the witch, such as the incubus and the familiar spirit, and yet historically it does not appear to have influenced these later ideas. Some of these elements probably are derived from Irish folk mythology (Russell 1972: 189–92; Russell 1981: 90–2). However, as we have seen in our discussion of witchcraft in other societies, cross-culturally the witch-belief has many common elements because societies share many basic values and common concerns.

Maleficium into Diabolic Witchcraft

By the end of the fourteenth century, *maleficium*, and sorcery in general, increasingly were being identified with heresy through the idea of the demonic pact. At the same time persecution of village 'witches' was increasing, along with that of heretics and of minorities such as Jews. Persecution probably was stimulated by the increasing social and economic instability of the late medieval period, as people and authorities sought personalized causes for the social and individual problems that this created. *Maleficium* was a secular offence, becoming increasingly identified with heresy, and secular, episcopal, and Inquisitorial courts were all concerned with the prosecution of witches. Consequently, all were involved in the final association of village *malefici* with heresy.

This was achieved through the application of inquisitorial procedure in accusations of witchcraft by all three courts. This procedure was weighted heavily against the accused. Its main feature was the use of torture to obtain confessions. It depended heavily upon denunciations, and penalties were rescinded for accusers who failed to prove their charge. The accused was rarely allowed a defence. Confessions had to be confirmed as 'voluntary' by the prisoner, which meant that torture was suspended to allow her to confess – under threat of further torture if she refused. The accused who withdrew her confession was sentenced to be burned alive.[16] From the fifteenth century, Continental witches were treated even more severely than other heretics and were burned on the first confession instead of upon relapse (Cohn 1976: 24; Monter 1976: 25–6; Russell 1972: 202–3; Russell 1981: 70, 77–8).

Fourteenth-century witch-trials reveal how courts, through use of the inquisitorial procedure, transformed charges of *maleficia* into confessions of diabolic witchcraft. Clerics, judges, and lawyers had developed the concept of diabolic witchcraft, and their beliefs guided the interrogation of the accused. Beliefs and procedures were locked in a vicious cycle of mutual reinforcement and development. The following example from the end of the fourteenth century is from Paris. The interrogation and trial were carried out by the secular authorities, but the proceedings were guided by the same beliefs as those held by papal inquisitors, and the same procedures were used.

Case IX-4

The trial began on 29 October 1390 at the provost's court. Accused were two women, Jehanne de Brigue and Macète de Ruilly. The indictments were of simple *maleficia*. It was alleged that Macète had employed Jehanne, who had a reputation for magic, to bewitch Hennequin de Ruilly into marrying her. The magic succeeded. Then Macète became dissatisfied with her husband and again employed Jehanne, who used toads and wax images to make him gravely ill.

The court believed diabolical practices must have been involved. They

---

16. See the 1438 case of Pierre Vallin – Russell 1972: 254–60; Russell 1981: 78–9. For two cases of torture in Lorraine in the Early Modern period, with contrasting results, see Briggs 1989: 7–8.

tortured Jehanne until she confessed that her grandmother taught her how to divine by calling up a demon named Haussibut. In order to do so she had to refrain from crossing herself or using holy water. She had been asked to make a pact with the demon by offering it one of her arms and a finger, to be collected on her death, but she said that she refused. In her testimony Macète claimed that only Jehanne had performed the magic against her husband. But she was tortured until she confessed that she had called up Lucifer for assistance.

Both women appealed to the *Parlement* of Paris against their convictions, but the *Parlement* upheld the verdicts of the lower court and they were burned on 19 August 1391 (Russell 1972: 214–15).

Through the operations of inquisitional procedures, inquisitors' beliefs and suspects' confessions were made to support and reinforce each other, and the number of witch trials increased, particularly in France, Germany, and the Alpine region. In the final quarter of the fifteenth century the Dominican Heinreich Institoris, Inquisitor for southern Germany, and his colleague, Jacob Sprenger, obtained the support of Pope Innocent VIII for their pursuit of witches in the Alps, which was being opposed by authorities unconvinced that witchcraft was extensive in Germany. In 1484 the pope produced a bull, *Summis desiderantes affectibus*, that firmly placed the seal of papal approval on the inquisitors' hunt for witches and finally ended the influence of the *Canon Episcopi* (Lea 1957: 226, 304–5; Robbins 1959: 263–6; Russell 1972: 229–30; Russell 1981: 78–9).

In 1486 Institoris and Sprenger produced a manual for witch-hunters entitled *Malleus Maleficarum* ('The Hammer of the Witches'), with the papal bull as its preface. Although modelled on previous inquisitorial handbooks, it became the most popular and influential inquisitors' manual.[17] It was reprinted at least thirteen times by 1520, and another sixteen times between 1574 and 1669. In part its influence was due to the development of printing; but it was mainly due to its argument for the link between *maleficia* and heresy. Witchcraft, it stated, was immediate and direct treason against God. As such, it was the most heinous of crimes and the most abominable of heresies. Because of its nature, any witnesses whatsoever were entitled to testify against a suspect, including excommunicates, convicted perjurers, and other criminals, and their identities could be withheld from the accused. Witchcraft had four essential characteristics: renunciation of the Christian Faith; devotion of body and soul to the performance of evil; the offering of unbaptized

---

17. Norman Cohn claims that its influence has been exaggerated, as the real witch 'holocausts' did not occur until the end of the sixteenth century (Cohn 1976: 225).

children to the Devil; and sexual orgies that included intercourse with the Devil. Witches could fly through the air and change their form. They abused the sacraments. They used magical ointments. The *Malleus* advocated stripping and shaving a witch to search for magical charms hidden on her body, but it made no mention of the Devil's mark or of some other important elements of sixteenth- and seventeenth-century witch-beliefs, such as the sabbat and the *osculum infame*. However, these were being discussed by other theologians. The *Malleus* also firmly fixed the association of witches with the female sex. (The grammar of its title implies that they are feminine.) Its explanation for this was that the female character is weaker, more stupid, and more carnal than the male (Anglo 1977a; Lea 1957: 306–36; Monter 1976: 24–6; Robbins 1959: 337–40; Russell 1972: 230–3; Russell 1981: 78–9).

The European conception of diabolic witchcraft that developed in late medieval times and provided the ideological justification for the witch-hunts of the sixteenth and seventeenth centuries was created by theologians and by the courts. A particularly important strand in its development was the stigmatizing of heretics as Devil-worshippers. This was the consequence of the medieval Church's seeking to consolidate and maintain its power and eradicate opposition to itself. Chapter 8 demonstrated that when witchcraft accusations are associated with positions of real power and wealth, witchcraft is treated as a very serious offence. In medieval Europe, powerful institutions and individuals used similar accusations against rivals and in pursuit of their interests. As a result, 'traditional' witchcraft became incorporated into a more complex conception of the witch that became part of mainstream Christian theology. Simply to be a witch became a capital offence. At the same time, witchcraft ceased to be merely an evil aimed against persons. It was treated as an attack upon society, because it was believed fundamentally to be opposed to Christianity.

–10–

'The Great Witch-Craze'

Europe's new image of the diabolical witch was almost complete by the middle of the fifteenth century (Cohn 1976: 225; Russell 1981: 76). It was created by churchmen, magistrates and lawyers, and was the view of witchcraft held by the educated classes (Briggs 1989: 16–20, 28; Cohn 1976: ix; Kieckhefer 1976: 4, 8; Larner 1981: 15, 23–4). Once formed, the concept of the diabolical witch remained virtually unchanged in Europe for almost 300 years (Larner 1985: 36; Russell 1981: 72, 86).

The popular conception of witchcraft continued to be concerned with *maleficium*, with injury to persons and property, and with relationships in the local community (Briggs 1989: 7–65; Ladurie 1987a: 5–78), and not with the witch as an agent of the Devil. These two worlds, the world of the educated and the world of the peasant and the labourer, with their different conceptions and concerns of witchcraft, met in the pursuit of the witch in the courts and through the legal process, and through this there developed some diffusion of beliefs and practices. Legal officers often sought evidence of *maleficium* as secondary evidence of witchcraft. Peasants seeking the cause of local misfortunes looked for the witch's mark on the bodies of suspects to confirm their suspicions (Briggs 1989: 18; Larner 1981: 201; Larner 1985: 32, 43–56, 74–6: Macfarlane 1985: ix).[1] Once suspects were formally accused the accusations were moulded and modified by magistrates and clergymen to conform to the Satanic pact (Holmes 1993: see in particular 70–1, Michael Dalton and the siting of the witch's mark).

The educated classes provided the inquisitors, lawyers and judges who investigated suspects. The majority appear to have been sincere individuals who believed in the rightness of their actions, notwithstanding the fact that a substantial portion of a convicted witch's property might be forfeit to the authorities (Russell 1981: 79; cf. Robbins 1959: 15–16; and see for example the case of Pierre Vallin – Russell 1972: 254–60; Russell 1981: 78–9). However, in Continental Europe the use of inquisitorial procedures produced confessions in which the accused had

---

1. For the relationship between popular and educated ideas of witchcraft, and the development of this relationship in late medieval times, see Kieckhefer 1976.

to name those with whom they consorted at the sabbat. (Their confessions constituted the main evidence, against themselves and against any persons they named.) In this way a legal investigation tended to create a widening network of suspects.

From the middle of the fifteenth century, witch-hunting increased dramatically, and generated local witch purges over a period of about two hundred years (Russell 1981: 72). This period has been called the European Witch-Craze. It had been estimated that some 500,000 persons were burned as witches (see, for example, Harris 1977: 146, 158), but recent examinations suggest that this figure is considerably exaggerated, and have led to its being much reduced. The estimate for Bamberg for 1624–31, for example, an area of the worst persecutions, has been revised from 600 down to nearer 300. English executions for the whole period were probably less than 500, and Larner revises the figure for Scotland, a country with high rates of prosecutions and executions, down to slightly over 1,000 from previous estimates of between 3,000 and 30,000 (Ladurie 1987a: 6–7; Larner 1985: 4, 36–7, 39, 71–2). A careful examination of France, one of the countries usually considered to have been in the forefront of witch-persecutions, suggests that executions may not have been significantly higher than in England, and indeed that the ratio of those executed to the total population may have been lower. As in England, there were marked regional variations in the amount of witch-hunting. In France, the centres of persecution were the eastern and northern frontier areas and the extreme southwest (Briggs 1989: 10–14; Soman 1986). It is incorrect to think of a 'witch-craze' sweeping across Europe. Instead, there were sporadic, local outbreaks of witch-persecution. However, Russell suggests that in Europe during the period of the 'Craze' 'at least 100,000 persons were executed' (Russell 1981: 11).

Large-scale witch-hunting began abruptly in the late fifteenth century, in northern Italy and southern Germany, and spread widely during the sixteenth century. The regions believed to have been most affected were France (but see the comment above), Germany, Switzerland, the Low Countries and northern Italy. These were areas where medieval heresies had been strong. In the sixteenth and seventeenth centuries witch-hunting spread into Scandinavia. It is usually claimed that, in order of severity, the strongest persecutions probably occurred in Germany, Scotland, France (however, see above), and Switzerland. There were serious isolated incidents in Holland, Sweden, the Basque country, and Puritan New England. These were countries in which the authorities were concerned with witchcraft as a diabolic activity. In England and Russia, where the emphasis was upon the personal harm committed by witches rather than upon witchcraft as a diabolic crime, there were milder persecutions. But even in those countries now considered notorious for the Witch-Craze, purges were episodic and very localized, and many areas escaped witch-hunting altogether. In areas like Spain (with the noted exception of the Basque country) and southern Italy there were almost no witch-hunts (Larner 1985: 3–4, 39–44, 88; Russell 1972: 268–9; Russell 1981: 76).[2]

Some historians see the Witch-Craze as being, at least in part, the result of the diabolical witch-belief's having developed a momentum of its own. They may place a particular blame upon Heinreich Institoris and the *Malleus Maleficarum* (Cohn 1976: 257; Larner 1981: 19–27; Monter 1976: 24–5; Russell 1981: 94; Trevor-Roper 1978: 12). However, they usually also see the Craze as the product of the Renaissance and the Reformation, and relate it to the social conditions that brought about these momentous events and to the social developments that they generated. The diabolical witch-belief had been constructed by Roman Catholic theologians, but it was accepted by Protestant reformers who rejected much of Catholic ritual and dogma (Trevor-Roper 1988: 115–18). In his classic examination of the Craze, Trevor-Roper ties it to the religious wars of the Reformation and the Counter-Reformation. In any region, the ascendant denomination applied the witch-image to its religious opponents (Trevor-Roper 1988: 64–7, 88–9, 112–16). Russell argues that the increase in witch prosecutions by Christians of all denominations in the sixteenth century was the result of the tensions generated by the religious conflict, wars, and associated population movements created by the Reformation. The height of the Craze occurred between 1560 and 1660, and was caused largely by the religious conflicts that culminated in the Thirty Years' War (Russell 1981: 82–3, 85).

The greatest persecutions were in the decades around the turn of the sixteenth and seventeenth centuries, a time when real incomes dropped to their lowest in the millennium. To survive, many persons were forced to sell their land and possessions and consequently suffered drastic loss of status. Europe was subjected to waves of impoverishment, economic stagnation, and general crisis. It is probable that these socio-economic factors, as well as the more specifically political ones, played an important role in promoting the tensions that fostered witch-persecutions (Briggs 1989: 59; Ladurie 1987a: 6).

Larner develops the relationship between religion and the Craze and relates it to the rise of the post-medieval state. She argues that, in the Early Modern period, Christianity became the first political ideology. It was used to legitimize the regimes of the new, more centralized and secular

---

2. For an outline chronology of witch-purges see Larner 1981: 16–19; Larner 1985: 39–44.

states of post-medieval Europe. In order to show their independence of the Papacy, regimes secularized areas of ecclesiastical law, including witchcraft,[3] and many of them adopted reformed versions of Christianity. For the governing classes, their subjects' allegiance to the adopted form of Christianity affirmed their acceptance of the legitimacy of the regime. It validated the dominance of governing classes, and strengthened their control by establishing ideological ties with their subjects. Religious deviance challenged both state and regime. Pursuit of witches and other nonconformists demonstrated both the regime's legitimacy and its control over the state and the processes of law and order. As a conflict theorist, believing that the development of witchcraft beliefs and accusations is likely to be related to issues of power, I find Larner's explanation appealing (Larner 1981: 192–7; Larner 1985: 64–6, 89, 90, 127–8, 134–9).

Throughout our examination we have seen that the witch is perceived as the most extreme type of deviant, the epitome of evil, perverting all that is of human or social value. In Early Modern Europe she had voluntarily dedicated herself to the Devil, and set herself against God, community, Church, state and regime. Consequently witchcraft was pursued with a particular ferocity. Taking action against witches was an especially effective way for the representatives of the state to demonstrate their power and commitment, and the strengthening of legal systems that was taking place in Early Modern Europe helped them to do so (Briggs 1989: 21, 59; Larner 1981: 195).

In England, for example, the first statute against witchcraft was enacted in 1542, towards the end of the reign of Henry VIII. It was repealed in 1547, apparently without the death penalty ever being applied. In 1563, the fifth year of Elizabeth's reign, a new statute was passed, with death as the penalty for murder by witchcraft and for a second conviction of causing bodily harm. A second conviction for destruction of property merited confiscation of the offender's property, as did the use of witchcraft for procuring love. A year's imprisonment with exposure in the pillory was the penalty for some less serious offences. In 1604, the second year of the reign of James I, a more severe act was passed. It punished by death those offences previously punished by life imprisonment. Although this act was influenced by Continental witchcraft doctrine, it retained the emphasis on witchcraft as an antisocial action, and not as a heresy (Macfarlane 1970a: 14–17; Peel and Southern 1969: 51–5; Thomas 1973: 225–7).

---

3. Not all states secularized witchcraft. In Spain and Mediterranean Europe, the Inquisition had sole control of witchcraft trials until the nineteenth century (Monter 1976: 25–6).

'The Great Witch-Craze'

Because of its secret nature, witchcraft, like poisoning, was treated as a special crime. Normal procedures of evidence were regarded as inadequate. Trials allowed evidence concerning matters that occurred years before the act with which the suspect was charged, and which were not directly related to it. The accused was not allowed to prepare a defence. Hearsay evidence was permitted, as was the evidence of children – including the suspect's own – and of hostile neighbours. Confessions of supposed accomplices were regarded as particularly important evidence. Stress was laid not only upon the character and actions of the accused, but also upon those of their parents and of their friends and associates. Because witchcraft was an occult activity, proven absence of the accused from the scene of the offence was considered irrelevant (Peel and Southern 1969: 57).

In attacking witches the actions of the authorities of the State coincided with the wishes of Church and populace. To Christian Churches, witchcraft was apostasy. To the governing classes, it was also subversion against the State. To the peasant, it threatened everyday life. The State in Early Modern Europe was more centralized and could pursue witchcraft and other forms of deviance more effectively (Larner 1985: 64–6, 89, 90, 127–8, 134–6).[4]

However, given that these conditions were widespread in Early Modern Europe, it is necessary to explain the local and sporadic nature of witch-crazes. Larner suggests that warfare with an external enemy or the presence of strong, organized heretical groups would militate against the development of witch-crazes. They constituted overt, politically organized enemies, and opposition to them created a focus for internal unity and the legitimacy of the regime. Regions that escaped witch-hunting included areas under armies of occupation, where rule was sanctioned by force and not by legitimacy. Similarly, the presence of ethnic minorities provided alternative groups to pursue. On the other hand, internal social unrest might promote witch-hunts as a consequence of a regime's attempt to legitimize its claim to power. Once a witch-hunt had occurred in any region it provided a precedent, and further ones were likely (Briggs 1989: 63; Larner 1981: 197; Larner 1985: 57–8). Robin Briggs suggests that idiosyncratic factors such as the presence of a particularly determined witch-hunter or epidemic diseases or exceptionally bad weather could promote sufficient denunciations to begin a craze (Briggs 1989: 63). A variety of factors could combine to create a local craze. The diabolical conception of witchcraft leads to a widening of accusations once suspects have confessed, and the most

---

4. Prosecution rates for some other forms of deviance, such as sexual offences, tended to mirror the rise and fall in rates of witchcraft prosecutions (Larner 1985: 59–63).

intensive witch-hunts occurred where the system of Roman Law was efficiently organized, with inquisitorial procedures condoning torture and guided by the Canon Law on witchcraft (Larner 1981: 197; Larner 1985: 4, 31–2, 38–9).

Larner suggests that a particularly important factor may be the *political* nature of the earliest accusations in a region. Witch trials involving important figures as victims or suspects tended to encourage the first outbreaks of witch-hunting. The use of accusations by elements of the governing classes for political purposes might cause the peasants to become aware that witchcraft beliefs were acceptable to the authorities, and make them more prepared to make formal accusations. Spectacular trials of a political nature might also arouse local fears of witchcraft. Larner suggests that the political use of witchcraft accusations could be 'a normal early stage in the growth of witch-hunting' (Larner 1985: 40–1). She notes the case of James VI of Scotland and the North Berwick witches. Before the Scottish witch trials of 1590–1, there was no serious attempt in Scotland to pursue witches, and the 1563 Statute, which prescribed death for witches and their clients, was not rigorously applied. After the trials there was a marked change in the beliefs and attitudes of the educated classes. James VI, who came to consider himself an expert on witchcraft (Clark 1977; Robbins 1959: 277–81), previously had shown little interest in the subject. His change of attitude in 1590 reflects the influence of the Danish court, where there was considerable interest in the diabolic idea of witchcraft. James took his queen from the court, and resided there for several months in 1589–90. What gave the 1590–1 witch trials their drama and stimulated Scottish fears about witchcraft was the claim that they concerned a conspiracy upon the life of the king, and that this was orchestrated by his political rival, the Earl of Bothwell (Larner 1985: 3–22).

Case X-1

The episode of the North Berwick witches began with Gelie Duncan, a servant girl with a reputation for healing. This aroused the suspicion of her master, David Seaton, the deputy bailiff of Trenant. He had her tortured until she confessed to being a witch and named others as the witches with whom she associated. One of these, Agnes Sampson, well-educated and elderly, a midwife and a healer of good reputation, was examined by the king and council and tortured until she confessed to 53 indictments of witchcraft. Among her confessions, she said that the Devil had given her foreknowledge of the storms of 1589 that prevented the king's bride, Anne of Denmark, from coming to Scotland to marry him. Another of those denounced by Gelie Duncan, Dr John Fian, the schoolmaster of Saltpans, confessed that he had taken part in actions to raise the 1589 storms. Agnes Sampson and John Fian were condemned, and were strangled and burned in January 1591.

As the trials progressed, it was claimed that a conspiracy had taken place against the life of the king. It was claimed that on Lammas E'en (31 July) 1590 a meeting of witches had taken place at Newhaven, presided over by the Devil in the form of a black man. Agnes Sampson proposed that they kill the king. Satan promised his help, but predicted that they would fail. He ordered them to make a concoction of toad, adderskin, and other ingredients with which to poison the king. At his instruction Agnes Sampson made a wax image of the king. She gave it to Satan to be enchanted, and he promised to return it at the next meeting to Barbara Napier and Euphemia MacCalzean for roasting. On All Hallows E'en 1590, seven score witches met at North Berwick kirk, including Agnes Sampson, Dr Fian, Barbara Napier, and Euphemia MacCalzean. The Devil appeared as a black man. He preached from the pulpit, calling upon them to be his faithful servants and promising that he would be a good master and they would never want. Some dissatisfaction was expressed that he had not yet enchanted the wax image given him on Lammas E'en, and he promised to do so. (Throughout this case the witches are presented as far from being supine servants of their diabolic master. Frequently they criticize him to his face.) Corpses were exhumed and dissected and their joints taken to be used as charms, and the assembly paid the Devil homage.

By the time this story had been assembled, Agnes Sampson and Dr Fian were dead. Barbara Napier, sister-in-law of the Lord of Carschoggill, wife of the burgess Archibald Douglas and an Edinburgh woman of some standing, was a known partisan of the Earl of Bothwell. She was acquitted of conspiring to kill the king, to James' great displeasure, but found guilty of other charges and sentenced to death. She pleaded her pregnancy, and it appears that some time later she was set free. Euphemia MacCalzean was also a prominent woman, the wife of Patrick Moscoop, a wealthy and influential man, and daughter of Lord Cliftonhall. After James had made clear his displeasure over the acquittal of Barbara Napier on the conspiracy charge, she was found guilty on nine charges, including the treasonable ones of attending the Newhaven and North Berwick conventicles. She was a friend of the Earl of Bothwell and a Roman Catholic, and James saw to it that she was burned without the mercy of strangulation. She confessed nothing.

The prosecution claimed that the witches' conspiracy had been at the behest of the king's political rival, Francis Stewart Hepburn, the powerful Earl of Bothwell. The main witness against Bothwell was a notorious wizard, Richie Graham, who gave evidence that Bothwell had sought his help to kill the king by witchcraft. James appears to have believed that Bothwell had indeed sought to destroy him by witchcraft. The Earl was arrested in April 1591. He admitted knowing Graham, but arrogantly declared his innocence of all charges and claimed they were concocted by his political enemies. In particular he accused the Scottish Chancellor, Maitland, and the English court, which distrusted him because of his reaction to the execution of Queen Mary Stewart. He claimed that Richie Graham was being bribed by the king and Maitland to testify against him, in the hope that his life would be spared. Certainly, Graham was well cared for in prison. (However, in February 1592 he was burned, perhaps in an attempt by James to appease the ministers of the Kirk, who were angered by his failure to prosecute the murderers of the Earl of Murray.) On 21 June 1591 Bothwell escaped. He was outlawed as a traitor and declared suspect of 'nicromancie and witchcraft [sic]' (Stafford 1953: 113). In July 1593 he seized control of the king and the court. His dominance was short-lived, and in 1595 he was forced to flee the realm, but while he was in control he contrived his own trial on the 1590–1 accusations and had himself acquitted.

So ended the trials of the North Berwick witches. Between November 1590 and May 1591 more than 100 suspects were examined, and a large, but unknown, number executed. The experience influenced James to write his *Daemonologie* in 1597 as a response to the Englishman Reginald Scot's attack on witch-hunting (*The Discoverie of Witchcraft*). The trials influenced the attitudes of the clergy, gentry, and legal profession towards witchcraft and gave encouragement to existing village attitudes towards witches. By 1597, when James revoked the standing commissions he had established to hunt out witches in Scotland, concern with witchcraft had been firmly established in Scottish society (Larner 1985: 3–22; Robbins 1959: 359–61; Stafford 1953).

English Witchcraft Trials

In England, although Continental beliefs influenced ideas about the nature of witchcraft, particularly those held by intellectuals and theologians, they did not alter the fundamental attitude towards witchcraft. It continued to be treated primarily as a relationship between individuals – an attack by one person upon another – rather than as a relationship between an individual and the Devil. Consequently throughout the period when witchcraft was a criminal offence it remained a felony that could be punished by hanging, and not by burning as was the case with heresy.

Persons accused of maleficent witchcraft, and against whom the magistrate considered there was a case, were tried by jury before the secular courts. Evidence considered sufficient to warrant a prosecution included: a bad reputation as a witch; cursing followed by injury of the one cursed, or known malice followed by injury of the person against whom malice was held; and blood relationship or friendship with a proven witch. Accusation by a 'white witch' or being named in a deathbed accusation constituted stronger evidence, as did accusation by a proven witch or possessing an 'unnatural' mark on the body. Torture of the kind used on the Continent was not allowed in English witchcraft trials unless the accused was also charged with treason, when those convicted were burned and not hanged. But attempts might be made to extort a confession by depriving the accused of sleep, starving her, feeding her only bread and water, beating, or maltreating her in other ways (Macfarlane 1970a: 18–20; Robbins 1959: 165; Thomas 1973: 617–18).

The preponderance of English witchcraft trials were in the second half of the sixteenth century and the first three quarters of the seventeenth, with the highest concentration in Elizabethan times. The total number of executions during this period was probably less than one thousand. However, witchcraft formed a substantial proportion of criminal proceedings, and this is only a partial reflection of local concern. Acquittal rates were very high, particularly in comparison with some Continental countries. The last witch hanged in England was in Exeter in 1685. The last to be condemned was in Huntingford in 1712. She was reprieved. The 1604 Act was repealed in 1736, but it had been controversial and rarely used for many years (Macfarlane 1970a: 95; Thomas 1970: 73; Thomas 1973: 534–41).

Before this period English witchcraft trials are rare, despite the existence of witchcraft beliefs of the kind found in later times. Thomas and Macfarlane argue that this is because the kinds of conflict that evoked witchcraft accusations, the conflicts between neighbours over charity that characterized the Early Modern period, had yet to be created. Given the widespread belief in witches and the way the witch-belief operates in other societies, I suggest that it is in part because fewer accusations reached the authorities, because most were treated as a purely local matter. In Tudor times, the formulation of witchcraft statutes encouraged formal accusations. Thomas and Macfarlane also stress the role of the Church in the Middle Ages in providing protection against witchcraft through prayers, ritual, and charms. This 'protection' largely was removed by the Reformation, while at the same time Early Modern England was a world of social changes that created new tensions between neighbours. A greater incidence of witchcraft fears between neighbours corresponded with the time when some of the most respected protections against witchcraft were withdrawn. These factors combined to produce the increase in formal accusations (Macfarlane 1970a: 130; Thomas 1973: 588–93).

Although English witchcraft trials placed little emphasis on a pact with the Devil or the witches' sabbat, intellectuals and theologians often were influenced by Continental ideas, and the wider public was acquainted with some Continental beliefs. The most famous witch prosecutions in English history, the Essex witch-finding movement of 1645, are the ones that show most strongly the influence of Continental ideas of witchcraft.

Case X-2

Matthew Hopkins was the son of a Suffolk minister. After practising unsuccessfully as a lawyer in Ipswich he moved to the village of Manningtree in Essex. This was the time of the English Civil Wars, and the eastern counties of England were the major region of support for Parliament. The short period (1645–6) during which Hopkins acquired notoriety as a witch-finder was a time of uncertainty and anxiety, when the outcome of the Civil Wars was still far from sure.

At Manningtree Hopkins accused Elizabeth Clarke, a one-legged old woman, of witchcraft, claiming he had heard a known witch identify her as one of her kind. She was arrested, and when other people came forward with further accusations Hopkins had her denied sleep for three days and nights until she confessed. She said she had been copulating with the Devil, who took the form of a gentleman with a laced band, for the past six or seven years. Hopkins and his fellow witch-finder, John Stearne, testified that they had seen her 'imps' (familiars). They included a white cat, a white dog, a polecat, a black imp, and a greyhound called Vinegar Tom. Their testimony was supported by six other persons, and Elizabeth Clarke was hanged for keeping evil spirits. She named six other witches, and the accusations widened until some 23 women had been charged. They were tried at Chelmsford on 29 July 1645 before Robert Rich, Earl of Warwick, and Sir Harbottle Grimston, and nineteen were hanged. The most common charge was murder by witchcraft, but nine were charged solely on the count of keeping evil spirits. Of these, seven were hanged. Hopkins later claimed that seven or eight of the witches used to meet in his neighbourhood every sixth Friday to sacrifice to the Devil. He said they were afraid of him and sent a bear to kill him in his garden.

Calling himself 'Witchfinder General', Hopkins now set himself up as a witch-finder. To guide his examination of suspects, he used King James' *Daemonologie*, Potts' book on the Lancashire Witches of 1612, and Richard Bernard's *Advice to Grand Jurymen*. He placed special importance on King James' writing on the keeping of familiars and regarded witchcraft principally as the keeping of 'imps', whose existence was revealed by the presence of marks on the witch's body. Suspects were stripped and searched for these marks.

It is said that Hopkins received a special commission from the Long

Parliament to carry out his work. Before the Chelmsford trials were over he began visiting other villages and towns, sometimes on request, but more usually on his own initiative. He was accompanied by his associate John Stearne and a team of two men and two women who searched suspects for witches' marks and suspected teats, and pricked them for insensible spots. Their activities extended from Essex into Norfolk and Suffolk, Cambridgeshire, Northamptonshire, Huntingdonshire and Bedfordshire. Local people made the accusations and the witch-finders examined the suspects. Because of the unreliability of records, it is not known how many persons were executed as a result of Hopkins' activities. John Stearne claimed there were some 200, but this appears to have been an exaggeration. However, between them Hopkins and Stearne must have caused almost as many executions in fourteen months as are recorded in England for the preceding 160 years. As well as the Chelmsford hangings, there were eighteen persons executed at Bury St Edmunds and twenty-six in Norfolk. And there were many others. The great majority were old women.

Hopkins' methods were to force confessions through starving the accused and preventing them sleeping. They were walked ceaselessly. Hopkins claimed they were kept awake to make their imps come to them. In August 1645 Parliament investigated his procedures, and he was prohibited from using some of them. His earlier victims were subjected to the water ordeal, but the Parliamentary commissioners prohibited this test. (However, long-standing witch suspects sometimes came to him and requested the ordeal in order to prove their innocence (Macfarlane 1970a: 41).)[5]

One of his victims was John Lowes, the octogenarian pastor of Brandestown in Suffolk. Lowes was a cantankerous old man who was unpopular with his parishioners and twice had been indicted for witchcraft. He was accused again, and Hopkins administered the water ordeal. It showed him to be a witch, and he was kept without sleep and walked back and forth until he confessed that he had made a pact with the Devil and possessed several familiars. He admitted causing the deaths of many cattle, and said he made his yellow imp sink a ship off Harwich, with the loss of fourteen lives. When Lowes recovered he retracted his confession, but he was hanged on 27 August 1645. The confessions Hopkins obtained by these methods reveal his acceptance of Continental,

---

5. See Douglas' observation on the poison ordeal among the Lele in Chapter 4. The water ordeal consisted of binding the suspect's thumb to her big toe, and placing her in a body of water. If she floated and did not sink, she was guilty (Robbins 1959: 492–4). When Hopkins was prohibited using the ordeal this effectively ended its official use in English witchcraft cases, but it continued an important part of popular belief (Holmes 1993: 69).

A Deed without a Name

as well as English, ideas of witchcraft. As well as keeping familiars and performing *maleficia*, his witches attended sabbats and sacrificed to the Devil and had intercourse with him.

In the spring of 1646 Hopkins' witch-hunting activities in Huntingdonshire were attacked by the Reverend John Gaule, vicar of Great Staughton. Although Hopkins examined at least twenty persons, of whom several were executed, opinion was turning against him. Gaule wrote a pamphlet, *Select Cases of Conscience touching Witches and Witchcraft* (published in 1646), to expose Hopkins' methods. It reveals the concern about witchcraft prevalent in eastern England at that time. He writes:

> Every old woman with a wrinkled face, a furred brow, a hairy lip, a gobber tooth, a squint eye, a squeaking voice, or a scolding tongue, having a rugged coat on her back, a skull-cap on her head, a spindle in her hand, and a Dog or Cat by her side, is not only suspected, but pronounced a witch. Every new disease, notable accident, miracle of nature, rarity of art, nay and strange work and just judgement of God; is by them accounted for no other, but an act and effort of witchcraft. (Quoted in Davies 1947: 154.)

In Norfolk the judges of assize accused Hopkins of extorting confessions by cruelty and creating 'witches' for his own profit. Certainly he and Stearne received substantial payment for their activities. They retired, and sought to exonerate themselves.

In 1647 Hopkins died of consumption. We have seen that witch-finders often are themselves accused of practising witchcraft. While this does not appear to have happened to Hopkins in his lifetime, it is interesting to note that soon after his death the rumour arose that he had been drowned whilst being 'swum' in the water test as a suspected witch.

It has been claimed that Hopkins' witch-hunting career was an expression of the Presbyterian intolerance of his period. In fact Hopkins' denomination is not known, and although the eastern counties of England were the stronghold of support for the Presbyterian Parliament there was much opposition within the region towards Hopkins' activities. Also, that same Presbyterian Parliament sent a commission to curb some of his excesses. The reason for the general fear of witchcraft appears to lie in the anxieties caused by troubled and uncertain times, in a region with a strong tradition of witchcraft. Hopkins' career took place during the English Civil Wars before their outcome could be predicted, and these supplied the conditions conducive to the development of England's greatest witch-hunt (Davies 1947: 150–4; Kitteredge 1958: 271–2, 331; Macfarlane 1970a: 135–42; Nottenstein 1911: 164–205; Parrinder 1963: 97–9; Robbins 1959: 249–53; Russell 1981: 97–100).

The Victims of the Great Witch-Craze

The majority of the victims of the Great Witch-Craze were women of the type discussed in our examination of witchcraft in England at this time (Chapter 6). They were usually elderly, and often widowed. They were poor and of low status, but often were not among the poorest and lowliest of society (Ladurie 1987a: 6–7; Larner 1981: 90; Monter 1976: 115–28).[6] The 'typical' witch was likely to be quarrelsome or a scold, and was often assertive, argumentative, and ready to defend herself verbally (Larner 1981: 89–98). Members of certain female occupations, such as midwives, may have been especially at risk of accusation (Forbes 1966: 112–32; Kitteredge 1958: 114–15; Midelfort 1972: 172, 187, 195; but see Harley 1990 and Holmes 1993: 72 for a re-evaluation of the importance of midwife witches).[7] The *Malleus Maleficarum* repeatedly warns against midwife witches, and gives examples of midwives' bewitching mothers and killing newborn babies for witchcraft purposes (Forbes 1966: 117, 127). Witch-suspects often had a reputation for using magic, and many were 'white witches'. Many were from families with a reputation for witchcraft, and might have had close relatives who had been accused, and even executed, in the past (Briggs 1989: 60; Cohn 1976: 248–9; Harris 1977: 168; Ladurie 1987a: 6–7; Larner 1985: 50, 61–2, 72–3, 84–5, 87; Monter 1976: 196–200; Russell 1972: 274; Trevor-Roper 1978: 120).

Men were much more rarely accused. Those who were, were usually criminals or relatives of female suspects. In countries on the periphery of the Great Craze, such as England and Russia, males represented 5 per cent or less of those accused.

The accused were usually persons against whom suspicions and local accusations had built up over time and who had become labelled as witches within the community. However, in areas where the Continental conception of witches dominated prosecutions, the authorities sought to make them implicate their confederates. They were likely to be made to implicate other persons with a local reputation for witchcraft. They might also denounce practitioners of benevolent magic, because they knew that their occult powers were considered diabolical by ecclesiastical authorities. When the net was widened in this way, men as well as women were likely to be accused. In Scotland, for example, as in the Jura region and in southwestern Germany and Venice, approximately 20 per cent of

---

6. See Larner 1981, for an examination of Scottish witches and a comparison with Keith Thomas' data on English witches (Thomas 1970; Thomas 1973).

7. See Russell 1981: 83, for an example of an accused midwife. Monter notes that midwives rarely were accused in the Jura region, and suggests, I think unconvincingly, that as this is a dairy region concern about *maleficia* focused upon the health of cows rather than of children (Monter 1976: 126–7).

accused persons were men (Larner 1981: 89–102; Larner 1985: 73; Midelfort 1972: 172; 187, 195; Monter 1976: 120).

We have seen that the position of women in many societies is an ambiguous and/or socially marginal one. As a result, these societies have a predisposition towards identifying women with witchcraft unless, as among the Western Apache, there is some cultural belief that associates witchcraft with men. In Europe, the predisposition was strengthened by the patriarchal emphasis of Christianity and its anti-female bias. Christianity identified woman with the temptress, the carnal, and the material. In Early Modern Europe the prosecution of women increased. Increasing numbers of older women were prosecuted for witchcraft and healing, while younger women were prosecuted for infanticide and prostitution. Kieckhefer's examination of witchcraft trials of late medieval times suggests that before the Witch-Craze, the ages of those tried for witchcraft were more varied, and men constituted approximately one-third of those brought before the courts, although the figure was declining in the later fifteenth century (Kieckhefer 1976: 93–105).[8]

Before the Early Modern period, women were the legal responsibility of husbands or fathers. The regimes of the developing states of Europe reduced women's legal dependence, and at the same time created new offences as they established and demonstrated their legitimacy and control of the State. More women found themselves prosecuted for female-associated crimes. These legal changes may have increased the ambiguous status of women (Briggs 1989: 60; Ladurie 1987a: 6–7; Larner 1981: 92–4, 100–2; Larner 1985: 60–5, 85–7; Midelfort 1972: 182–6; Monter 1976: 197–8; Russell 1981: 113–18). Midelfort has suggested that the demographic changes that took place in Western Europe in the sixteenth century particularly affected the status of women, and made certain categories especially vulnerable to accusation. Changes in family patterns created a large group of spinsters and widows. The resulting social problems affected attitudes towards them, and consequently they constituted a high proportion of those accused of witchcraft (Midelfort 1972: 183–7, 195–6; Monter 1976: 123–4). Intellectual élites explained the preponderance of female accused by claiming that women were naturally inferior to men, and therefore more susceptible to Satanic temptation.[9]

---

8. For opposed assessments of the percentage of men accused of witchcraft in the Middle Ages prior to the Great Witch-Craze, compare Cohn 1976: 248; Russell 1972: 279.

9. For a detailed account of beliefs about women in witchcraft in Early Modern England see Holmes 1993. Popular explanation in Early Modern England for the predominance of female witches appears not to have stressed the idea of the inferiority of women, but to have related it to the manner in which witchcraft was transmitted

'The Great Witch-Craze'

In the famous Salem witch-hunt in colonial Massachusetts in 1692, at the end of the European Witch-Craze,[10] we still find operating many of the principles that characterize witchcraft in other societies. The first of the accused were from the antisocial and the misfits in the community – the pipe-smoking beggar woman, the West Indian slave, and the outsider with a scandalous sexual history. The accusations occurred at a time of social and economic insecurity, and of political instability. From 1688, with the deposing in England of James II and the ousting in Massachusetts of his governor, Sir Edmund Andros, the colony entered a period of political uncertainty.

At the same time, at the end of the seventeenth century, social and economic changes in the American colonies – and in the world – had resulted in the development of two factions in Salem Village – a traditionalist faction that sought the independence of Salem Village from the town of Salem, and a faction whose leaders represented a new, growing mercantile capitalism and wished the Village to remain part of the Town. However, the Puritan ethic involved an ideal of the community as a unified organic entity. The factional dispute was seen in moral terms, and opponents were represented implicitly as enemies of the community and servants of the Devil. The pattern of accusations was influenced strongly by the Village factionalism. Within the Village, the great majority of the accused and of their defenders were associated with the pro-Town faction, while the great majority of accusers belonged to the pro-Village faction. Although it appears unlikely that the behaviour of the accusing girls was cynically manipulated by the pro-Village leadership, who believed them to be afflicted by witchcraft, it was interpreted and directed by adult members of this faction.

In Salem Village, many of the accused were outsiders to the

(Holmes 1993: 51). Some writers claim the European Witch-Craze was a consequence of Early Modern European society's being 'women-hating' (see, for example, Ochshorn 1994). My argument is that the predominance of women is the consequence of the social position given to women. In Europe, their marginal and ambiguous position made them more likely to be suspects than men, and the more marginal, ambiguous and deprived a woman's position the greater the possibility of her being accused. (That women's marginality and deprivation is unjust and unmerited, I take for granted.) For a reasoned examination of the extent to which European witch-hunting was women-hunting, see Larner 1985: 79–91.

10. For interpretations of the Salem craze see, for example, Upham (1867), who ascribes the events to Village quarrelling. Starkey (1963) ascribes them to Puritan repression. Bednardski (1970) sees the craze as the consequence of the challenge to Puritan values by socio-economic and ideological changes of the time. Hansen (1971) believes there were people who practised malevolent magic at Salem, and that there is considerable evidence of white magic. Boyer and Nissenbaum (1974) blame the developing influence of mercantile capitalism. Robbins (1959), as usual, blames the sheer bloody-mindedness of the accusing girls.

community, and this was expressed in the way they had led their lives. Many were not only outsiders; their lives demonstrated the economic changes and vagaries from which the Villagers felt threatened. John Proctor, the most prominent Villager hanged in the Craze, was a native of Ipswich, Massachusetts, who had advanced himself socially and economically. Although he lived in Salem Village, he had less contact with its affairs than he had with Boston or Salem Town (Boyer and Nissenbaum 1974: 200–2). On the other hand Sarah Good, one of the first three accused, had come down from the child of well-to-do parents into poverty. She did not accept her condition gracefully, and this was expressed in her hostile attitude towards those from whom she sought charity (Boyer and Nissenbaum 1974: 203–6).

Rossell Hope Robbins, who adopts a strongly anticlerical interpretation of the Witch-Craze, argues that the main motivation behind witch-hunting was the desire of the authorities involved to appropriate the property of their victims (Robbins 1959: 15–16). However, the great majority of those convicted had little or no wealth, and most prosecutions were an expense to the community (Larner 1981: 197). Only in areas where the legal system did not operate to restrict the extension of accusations did individual crazes tend to get out of hand. Under these circumstances prominent individuals of substantial means might be accused and convicted (Monter 1976: 124–6; Russell 1981: 83–4: see this reference for the trial of Dietrich Flade, a judge, at Trier in 1589). In a detailed examination of confiscations in southwestern Germany between 1562 and 1684, Midelfort shows that, although most of those executed as witches were poor, a disproportionately large number of wealthy persons were accused. However, the proportion of a guilty person's property that was confiscated was not excessive. Usually it was easily payable, and excessive expropriations were likely to promote protests leading to their prohibition (Midelfort 1972: 164–78, 195). When persons of good reputation were accused it was likely that scepticism would be aroused and the witch-hunt would be brought to an end.

Midelfort demonstrates that, in the large panics of southwestern Germany, as accusations spread persons of higher social status and greater wealth were drawn in. He suggests that this is because they were well known, at least by reputation, and so were likely to be names that readily would come to the minds of people under torture. Consequently the numbers of wealthy accused tended to rise in the later phases of witch-hunts, as did the proportion of men. After the 1620s men became more common among those accused, as did children. Midelfort argues that this reflects the breaking down of the traditional stereotype of the witch. He suggests that the realization that no category of persons was now safe was an important factor influencing opinion against witch-hunting, ending public trials of large numbers of accused in this region, where the Witch-Craze had become most developed (Midelfort 1972: 72, 178–82, 190–2).

Case X-3

The extension of accusations to form a local craze can be illustrated by a small craze from the city of Neuchâtel, in the Jura mountains, in 1582–3. It followed an outbreak of plague. In January 1582 Jehanne Berna, a widow from Savoy, was arrested for theft and for having attended a mysterious meeting of several women that had been observed some weeks before. Under torture, she confessed to both charges, admitted that the meeting was a sabbat, and named three other women who she said had participated. (None were native to the county of Neuchâtel.) All three were arrested. Each added further details about the sabbat, and between them they named another twenty-four attenders, of whom four, all women native to the county, were arrested. Two of these, Madeleine Merlou and Anthonia Preudhon, confessed to attending the sabbat. Madeleine said the Devil gave them herbs with which to make grease to spread the plague. Both were condemned to be burned alive. Anthonia hanged herself in her cell, and her body was thrown on Madeleine's fire. A third accused escaped and tried to kill herself with a knife, but failed and was recaptured. She made a similar confession to the others and was burned in February. In the meantime the four original accused had been found guilty of witchcraft and burned.

Two more accused women confessed and were burned. Another, Marguerite Lambert, named by three of the 'witches', refused to confess in spite of her tortures, and was banished from Neuchâtel after agreeing not to harm anyone on account of her imprisonment. She was allowed to return the following year.

Here the 1582–3 craze ended. There had been eight executions, one suicide, and a banishment. Among those named in the confessions but not arrested were Blaise Cartier, widowed mother of the exiled Marguerite Lambert, and Jacqua Lainhard. Eighteen years later both were charged with witchcraft and executed. The fact that they had been denounced during the 1582–3 craze helped send them to the bonfire (Monter 1976: 92–5).

In the Jura region (in what is now western Switzerland and southeastern France) in the sixteenth and seventeenth centuries there were numerous local witch-crazes, but they were small in scale, with few victims. Many of those accused were never arrested, or were acquitted.

The rate of acquittals varied with different political regimes, as did the percentage of male accused. In those parts of the Jura that first equated witchcraft with heresy, almost 40 per cent of convicted witches were men – the highest recorded incidence in Europe. In French-speaking Switzerland the figure is about 20 per cent, similar to the average for southwest Germany, where the greatest witch-hunting took place (Monter 1976: 195–6).

The reason for the relative liberalism lies in the legal procedure used to examine witches. Although torture was used, it was within the restrictions laid down by the Imperial Law Code of 1532. Most of the women to whom it was applied, and about half the men, proved able to withstand it and did not confess. (However, in Case X-3, it proved sufficient to extend the craze, and to produce increasing elaboration of the details of the supposed 'sabbat'.) Also, judges paid little attention to accusations made by children. In addition, witchcraft examinations in this region were concerned with *maleficium* as well as with attendance at the sabbat. As in England, a person must be suspected of acts of *maleficium* before she would be arrested. (In many cases, this would mean that she already had a local reputation as a witch.) In particular, this bred caution in dealing with persons denounced in the second and third wave of confessions (Monter 1976: 7–12, 191–200).

Because of these constraints witch-crazes in the Jura never reached the proportions that they did in southwest Germany. As well as the lack of legal constraints, southwest Germany also suffered from religious conflict, and hundreds of persons might be executed as witches over a few years in small areas (Midelfort 1972; Trevor-Roper 1978: 120).

An Organization of Witches?

The persecution of witches in Europe from late medieval times, and the belief in an organized conspiracy of witches upon which it was based, have led some historians to argue that some kind of widespread cult of witches actually existed.[11] This idea was developed by nineteenth-century academics, the most renowned of whom was the great French social historian, Jules Michelet. In his famous work, *La Sorcière*,[12] he argued that witchcraft was a protest cult of medieval peasants and women.

The most influential twentieth-century exponent of this view was Margaret Murray (Murray 1921; Murray 1934). She popularized the idea that witchcraft in Western Europe was an organized religious cult, the remnant of an ancient fertility religion driven underground by Christianity.

---

11. For an account of the development of this idea, see Cohn 1976: 102–25.
12. Translated into English as *Satanism and Witchcraft* (Michelet 1960).

Murray's interpretation is not tenable. It is wildly speculative, and based upon selective doctoring of case material.[13] But some serious contemporary historians continue to argue that the witch persecutions were aimed at an organized form of esoteric religious activity. Probably the most eminent of these is Professor Jeffrey Russell (Cohn 1976: 121–5; Russell 1972: 3, 280–3), for whom the strongest evidence for the existence of a fertility cult associated with witchcraft is Carlo Ginzburg's account of the beliefs of the sixteenth- and seventeenth-century peasants of the Friuli region of northern Italy, and possibly of adjacent regions, concerning an anti-witch cult, the *benandanti* (Ginzburg 1966; Russell 1972: 41–2; Russell 1981: 76).

The peasants believed that the *benandanti* were men and women who went out on certain nights to fight witches. Both witches and *benandanti* were organized along military lines, with companies and captains. The *benandanti* were armed with sticks of fennel and the witches with stalks of sorghum. If the witches won, the crops would fail and there would be famine. If the *benandanti* won there would be a plentiful harvest. The *benandanti*, then, were perceived as a fertility cult. Its members were those persons, of either sex, born with a caul. They were called to service when they were twenty years old and served until about forty. Individual *benandante* could also identify witches and cure persons who were bewitched.

From Ginzburg's account it is quite clear that the organization and the battles were not real in any material sense. (He stresses this in his later repudiations of the interpretations of his analysis by Russell and by Midelfort (Ginzburg 1992: 10, 26 n. 38).) They existed in the dreams of individual *benandante*. A *benandante* participated in these activities 'in spirit' (Ginzburg 1966: 3, 17, 19, 58–9). We know that such dream battles against witches are not unique to the north Italian peasants of Early Modern Europe. We are back with the Nyakyusa and Safwa of Chapter 5, where witch fights anti-witch for control of the village's fields while their material bodies remain asleep at home. The fact that individual *benandante* held differing beliefs about the nature of their organization and its activities may be taken as an indication of their fictional nature.[14] Such beliefs do not demonstrate the existence of a witches' organization.

Similar beliefs appear to have been widespread in southeastern Europe. Gábor Klaniczay points to the similarities between the *benandanti* of Friuli and the *kresniks* of neighbouring Slovenia and Istria, who were born with a caul and were also believed to fight witches in their sleep. He suggests that the *kresniks* too may have been regarded as a fertility cult.

---

13. See for example, Cohn 1976: 107–15.
14. See Ginzburg 1966: 84 for an example.

A Deed without a Name

Both *benandanti* and *kresniks* may have been a development of a more archaic folk-belief concerning fights between clan sorcerers (*kresniks*) in defence of the fertility of their clan lands, and between sorcerers and werewolves. Again, these fights took place when the parties were asleep (Klaniczay 1990: 129–41).[15]

Between 1610 and 1640, under pressure from the Inquisition, beliefs about the *benandanti* became transformed to conform to the ecclesiastical conceptions of witchcraft. The *benandanti* became witches, and the night meetings the sabbat with the Devil. This transition was assisted in the popular mind by the fact that *benandante* could cure illness caused by witchcraft, and were able to identify witches. They were perceived as 'good witches'. If they had these abilities, it was easy to assume that they could abuse them and be witches themselves.

Even if they play down the idea of an organized fertility cult, some authorities claim that a degree of formal organization could exist between individuals committed to performing activities associated with witchcraft. They suggest that such activities were especially attractive to women. Russell argues that, because of their socially disadvantaged position, in medieval times women particularly were attracted to protest movements, and to ecstatic movements like the dance crazes. Their over-representation in protest movements and in heretical movements such as the Waldensians was an expression of rebellion against the male-dominated society.

But their rebelliousness could be expressed in a number of forms, one of the more covert being witchcraft. Devil-worshipping sects existed and performed some of the activities associated with the diabolical witch-image, such as killing and eating babies (Cohn 1976: 121–3; Russell 1972: 280–3; Russell 1981: 113–18). Other writers have also suggested there was some truth to the sabbat, which could be traced to the deprived position of peasants or the powerlessness of women (see, for example, Ginzburg 1966: 134–5).

However, the idea of a local witches' organization, or even a universal one, is widespread, although it is not as elaborate as in literate medieval and post-medieval European society. We have seen the local coven of the Navaho, with its officials whose positions are based upon their social standing in everyday life. There is the nightly gathering of powerful individuals from a Tiv district, who wish to increase their innate personal power. In his dream Don Talayesva attended a gathering of witches in their *kiva* beneath Red Cliff Mesa. The Hopi believe witches have a world-wide organization, containing people of every race and nation. Don believes Hopi witches are the worst of all and form the leaders of this

---

15. This reference contains an excellent summary of Ginzburg's account of the *benandanti*.

sect. A cultural belief in a secret organization of witches is no evidence of any real organization. At root, the witch-belief is concerned with problems of human existence, not with real organizations of evil-doers.

This is not to say that in societies with the witch-belief many individuals may not believe they possess the ability to influence events by supernatural means. The majority of cunning folk believed in their supposed powers, and some persons, including some of those victims of the European Witch-Craze who confessed to fantastic crimes, undoubtedly believed they had used their powers to harm others (Cohn 1976: 250; Kieckhefer 1976: 97–105; Larner 1981: 94–6; Monter 1976: 199–200).[16] But all this is a far cry from the physical existence of a widespread witches' organization.

The End of the Witch-Craze

The European Witch-Craze came to an end in the second half of the seventeenth century. There had always been sceptics who, often at considerable personal risk, were prepared to publish and speak out against the tide of opinion which was in favour of witch-hunting (Monter 1976: 32–5; Robbins 1959: 125–6; Russell 1981: 123–4; Trevor-Roper 1978: 73–5). They included men like the physician Johan Weyer (1518–88), who published his *De Praestigiis Daemonum* in 1563, which argued that old women who believed they had performed evil were suffering a hallucination inspired by the Devil (Klaniczay 1990: 175–6), and the English gentleman Reginald Scot (1538–99) whose *The Discoverie of Witchcraft*, published in 1581, argued that spiritual manifestations were impostures or illusions formed in the mind of the observer (Anglo 1977b; Clark 1977; Robbins 1959: 453–4).

The Witch-Craze ended when the attitudes of educated élites changed towards witchcraft as a crime. In the late seventeenth century the societies of Early Modern Europe had yet to undergo major changes in their social structures. They were still peasant economies. Although, after 1660, plague was vanishing, infant mortality remained very high and there were still famines. The advances in medicine and technology that would make life less precarious had yet to take place, and it remained as hazardous as at the height of the Witch-Craze. Witch-hunting ceased because judges and lawyers and juries, drawn from the educated classes, ceased to convict. They recognized the unreliability of evidence of witchcraft. At the same time a new philosophy was emerging that viewed the world as

---

16. See Muchembled 1981, in particular 42–3, for an examination of the psychological processes whereby an accused may come to believe that she is a witch. Ladurie (1987a: 18) has a précis of Muchembled's argument.

created by God to operate according to universal mechanical laws that could be discovered and described. If the causes of apparently 'miraculous' events were not known, the new, mechanical philosophy bred optimism that one day they would be exposed and understood. The educated élites, who controlled the courts and made the laws, ceased to believe in a world of immediate supernatural causes. The courts ceased to convict, and Witchcraft Acts were repealed (Briggs 1989: 22, 45–52, 59, 63; Ladurie 1987a: 14–17; Monter 1976: 26, 34–42; Russell 1981: 121–31; Thomas 1973: 681–98). In Britain a statute of 1736 repealed the existing witchcraft statutes of England and Scotland. It removed the offence of causing harm by witchcraft, but continued to allow the prosecution of persons who claimed magical powers, on the grounds that such claims were fraudulent.

Within this general explanation in terms of élites adopting a mechanistic, Cartesian philosophy, Christina Larner gives a more political explanation for the ending of witch prosecutions. She argues that the crucial factor was that Christianity ceased to be the legitimizing ideology of states and regimes. Educated élites developed new ideologies based upon secular concepts such as the pursuit of liberty, progress, enlightenment and patriotism. Consequently the prosecution of witches lost its political importance (Larner 1985: 66, 90, 139).

In France, the decline in successful witchcraft prosecutions began particularly early, and demonstrates the crucial role of educated élites. The development of the power of the French Crown resulted in a centralized judicial system, with appeals from local courts to higher ones. This system was dominated by the *Parlement* of Paris, whose judges had power and prestige unequalled in legal systems elsewhere in Europe. They were experienced full-time professionals, aware of the problems of evidence of witchcraft, and of the legal abuses that took place in lower courts. As early as the 1580s they began to restrain the actions of lower courts in witchcraft cases. The critical attitude of the Paris judges influenced provincial *parlements*. In 1624 automatic appeal to the Paris *Parlement* of sentences involving torture, corporal punishment, or execution was instituted, and it began dismissing large numbers of witchcraft cases long before the royal ordinance of 1682 did much to decriminalize witchcraft. The French developments contrast with neighbouring regions in the same period, such as Lorraine, Luxembourg, and the Spanish-speaking Netherlands, which lacked the kind of formal appeal system developed in France (Briggs 1989: 13–14, 31–2, 44–51, 59, 63; Ladurie 1987a: 16).[17]

---

17. For the 'standard' historical interpretation of Early Modern France, that contrasts with Briggs' revision, see Thomas 1973: 537.

Continuing Popular Beliefs in Witchcraft

In France, the parting between popular and élite attitudes towards witchcraft was influenced decisively by local witch-crazes in 1642–4, as the educated classes reacted to the spectacle of uncontrolled popular justice (Briggs 1989: 50–2). In Western Europe the crucial change in belief in witchcraft occurred among governing élites, and not the general populace. Popular beliefs in witchcraft continued in village communities well into the nineteenth century, and they remained strong in some relatively isolated areas into this century, along with traditional ideas about healing and magic. They were expressed in informal and illegal action against those suspected by the community. Peasants, distrusting courts and formal legal procedures, had always preferred to use less formal methods to deal with suspected witches, from protective magic to direct action, including lynching, and this may be one important reason why, although the legal machinery for prosecution was available in Early Modern Europe, there was in fact no continuous, widespread, Craze. Peasants were hesitant to use the courts. But once a deposition was made large numbers of persons often were prepared to testify. (This is another indication that persons formally accused often were those who had acquired a widespread reputation for witchcraft, built up over a long period of time (Briggs 1989: 22, 28–9, 37–44, 63–4).) Now, within a few decades, élite attitudes to witchcraft changed to outright disbelief and, with the ending of legal witch-prosecutions, an accuser could be prosecuted for slander (see, for example, the case of Gérarde Mimalé – Ladurie 1987a: 29).

Case X-4

Ruth Osborne and her husband lived at Tring in Hertfordshire. They were very old and very poor, and were reputed witches. In addition, they were supporters of the unpopular Jacobite rebellion of 1745. Ruth Osborne was believed to have bewitched a farmer, one Butterfield, because he refused to give her some buttermilk. He took fits and his cows died. Witchcraft was confirmed by a wise woman of Northampton. In 1751 a local rumour spread that suspected witches were to be put to the water ordeal in Longmarston. The Osbornes were taken into the workhouse for their protection, but on 22 April a mob broke in and dragged them, naked and bound, to a stream and threw them in. Mrs Osborne failed to sink. She was thrown on the river bank, where she quickly died from her mistreatment. Her dead body was kicked and beaten with sticks. Her husband was similarly treated and also died. The leader of the mob, Thomas Colley, a chimney-sweep, received a collection from its grateful members. The authorities took a different view. They charged him with wilful murder, and he was hanged on 21 August 1751 (Robbins 1959: 368–9).

In England witchcraft accusations had always been generated from the local community (as were the majority of accusations on the Continent (Briggs 1989: 7–65; Ladurie 1987a: 5–78)). However, changes in attitudes towards the poor, and new institutions for dealing with them, may have caused accusations to diminish by the end of the seventeenth century. By this time the conflict between charity and individualism was being resolved. The position of the poor had become legally regulated: the Poor Law had become a regular system of relief, and their support was no longer a moral obligation for their neighbours (Macfarlane 1970a: 204–6; Thomas 1973: 538–9, 681–97).

In Western Europe, local witch-beliefs gradually lost their importance as the lack of legal sanctions against suspects, and the threat of legal action against accusers, made it impossible to 'prove' witchcraft or take effective action against suspects. Consequently, witch-beliefs gradually were replaced within the community by other explanations for misfortune, that enabled other kinds of actions to be taken to remedy or prevent them (Monter 1976: 40; Thomas 1970: 69–73; Thomas 1973: 681–98). Nonetheless, witch-beliefs have often exhibited a remarkable degree of resilience.

Writing of witchcraft beliefs in the Basque country of Spain for the period between 1935 and 1950, Julio Caro Baroja says that witches have ceased to be linked with the Devil in the popular imagination. Instead, they are associated with goddesses of Antiquity and folk goddesses, such as Diana and Holda. Witches have the power of the evil eye and can cause sickness and other misfortunes, diagnosed by local diviners who may name the witch responsible. Witches hold nocturnal meetings at crossroads. Elsewhere in rural Spain all the witches of a region may have a common meeting place. Those of Galicia meet at Coiro; of Burgos at Ceregula (Baroja 1965: 227–40).

At the beginning of the twentieth century, Dr R. C. Maclagan recorded many accounts of 'evil eye' from the Western Highlands of Scotland (Maclagan 1972). Sudden, unexpected personal misfortunes often were attributed to the innate power of malign individuals. In particular, misfortune was effected by their gaze. This power was activated by feelings of envy or covetousness, and although it might operate unconsciously it was often believed to be used deliberately. Evil eye afflicted domestic animals, milk and butter, and people's health. Cattle were especially susceptible. Anything outstanding or exceptionally handsome was a likely target. Consequently people publicly belittled their property and parents tried to hide the beauty of their children, and praise by others could arouse suspicion.

Case X-5

'There was a woman in Islay who had a sow that had a litter of young pigs. One morning a neighbour came to her, and seeing the teapot at the fire, asked if she would give her a drop of the tea. This was given to her. She then said, "You have fine young pigs there." "Yes, but they are all sold," answered the other. With that the visitor turned towards the door to go, and said, "Well, you sold a pig to me that died." That was all that passed, and she went her way; but the following morning two of the pigs were dead, and the owner was quite certain that the woman had done the mischief. The reciter who gives the information is of the same opinion, and he adds that it was an unkind thing to do on a poor woman' (Maclagan 1972: 41–2).

Suspects were usually neighbours, especially if they were old women who were disliked because of their odd behaviour. However, strangers also were suspect. It was believed that both the evil eye and the power (*eolas*) to cure its effects, often were transmitted within families.

Many charms were used for protection from the evil eye. Cures for the misfortunes it caused were undertaken by persons with special knowledge about such matters. These *eolas* doctors[18] performed magic rituals to turn the evil back on the witch,[19] or to transfer it elsewhere. Usually, the *eolas* doctor helped identify the witch. Often he or she gave a vague description of the offender and left the client to put a name to it. The *eolas* doctor might question his clients to learn their suspicions, and sometimes the client was made to look into a magic mirror to see the witch. Many of the techniques and procedures used by *eolas* doctors in Gaelic-speaking Scotland to identify witches at the beginning of the twentieth century are reminiscent of Keith Thomas' and Alan Macfarlane's descriptions of the operations of cunning men and other diviners in Early Modern England (Maclagan 1972: 205–13; Macfarlane 1970a: 115–34; Thomas 1973: 219–20, 247, 315, 654–6).

Collecting cases of witchcraft from the first half of the twentieth century among the peasants of Franconian Switzerland (northern Bavaria), Hans Sebald (Sebald 1978) records that families often held

---

18. The term '*eolas* doctor' is my own.
19. This is my term. Maclagan simply describes them as persons with the evil eye; but by our definition they are witches.

A Deed without a Name

witchcraft responsible for sickness and death in animals and humans, and for many kinds of strange misfortunes. By the twentieth century the idea of the witches' sabbat and the witches' coven had been lost. (They had been prominent features of local beliefs in previous centuries, when it was held that the witches of this region met annually at the Blocksberg mountain.) The witch had come to be conceived as a more solitary figure who worked harm alone. There were several kinds of witch. The *milchhexe* milked cows by magic and stole their milk. The *stallhexe* attacked the stable by bringing sickness to cows, calves, and hogs and causing cows to give bloody milk. The *pferdehexe* attacked horses, and advertised her activities by braiding their tails. In order to cast a hex, the witch had to consult a *grimoire* – a manual of magic. This was a forbidden book, the *Sixth Book of Moses*.[20] She also had to obtain an item of property belonging to the victim, or introduce an item of her own property into the victim's home. Although it was believed that witchcraft ran in families and was passed particularly from mother to daughter, it was regarded as an act of free will because the witch had to consult the forbidden book.

The majority of witches were women (as were the majority of healers, who countered witchcraft). They were usually middle-aged to elderly women who were quarrelsome, self-centred and stingy, and were believed to obtain satisfaction from the misfortunes of others. They threatened their victims, often in ambiguous terms. Gradually such a person acquired a bad reputation as an occult practitioner, and her estrangement from her neighbours became strengthened. Victims always accused neighbours or relatives, although strangers were also suspect. Those they accused were persons with whom they had quarrelled, or who they believed were envious of them or hated them for some reason.

We began this examination of the witch with an example of impotence being blamed upon witchcraft, from the Willisau in 1531, the case of Stürmlin and Sebastian (Case II-1). It seems appropriate that our final example of traditional witchcraft in Europe should also be a case of impotence, from the Normandy Bocage in the 1960s. Jeanne Favret-Saada's sophisticated account of witchcraft beliefs among the peasants of the contemporary Bocage[21] reveals how secrecy and ambivalence are

---

20. There are parallels between Maclagan's and Sebald's accounts of witchcraft and Randolph's account of witchcraft in the Ozarks, USA (Randolph 1947: 120–61, 264–300).

21. Favret-Saada 1980. Favret-Saada's account is the most sophisticated and subtle analysis of modern witchcraft beliefs. Neither my account of witchcraft beliefs in the Bocage, nor my precis of the case study, can in any way reflect the detail and complexity of her analysis. In particular, my account of Jean Babin's case reduces it to a crude representation of the original. My excuse is that I have limited space, and that the simplified account that I give is adequate for my purpose. Interested persons are recommended to read the book.

essential to their functioning. People do not talk about witchcraft, for fear of ridicule. Many profess not to believe, and many do not believe until they are caught up by misfortunes they think cannot be explained by recourse to logic and science. These are misfortunes that are abnormally repetitive and recur over a long period of time. When people come to believe they are bewitched it is not only fear of ridicule that prevents them from making the matter public. If they talk about it outside the close circle of those associated with identifying and combating the witch, they leave themselves vulnerable to further attacks.

Witches behave oddly and possess supernatural power that operates through their words, touch, and glance. They are never enemies to each other, only accomplices. (This contrasts with Franconian Switzerland, for example, where one witch may be employed against another.) It is believed that witch families marry into one another, in order to increase their power. Even innocent members of such families are dangerous, as they are conduits through which their witch-relatives can operate. A witch's supernatural power is acquired through reading certain occult books. No one ever admits to being a witch, and a public accusation is never made.

When someone is subject to misfortunes that he cannot explain satisfactorily, someone else, a believer, usually takes it upon himself to suggest that the cause may be witchcraft. Witchcraft is aimed always at the head of the family, even when it attacks the person or property of another member. It is usually combated by an unwitcher, a person with the gift of power who diagnoses the client's condition. If it is witchcraft, and if the unwitcher believes his (or her) power is strong enough, he trades mystical blows, at a distance, with the suspect until one of them is vanquished. This is another reason why believers, who are usually persons who have been bewitched themselves, will not discuss witchcraft openly. In order to be cured they have sought the death of their witch, even if in the end he only suffered a diminishing of his power. For the evil to be removed, one must 'trade evil for evil'.

In contrast to the witch, an unwitcher acknowledges his ritual profession. He, or she, takes only symbolic payment for his, or her, services, but claims they will be effective only if the client believes the diagnosis. Because of this, and because the unwitcher risks his life on their behalf, clients should tell their unwitcher everything that they believe may be of relevance to their case. By the time a client consults an unwitcher, usually he is convinced that witchcraft is the cause of his misfortunes and sure of the identity of the guilty party. He prepares a list of names for the unwitcher, who guides him to pick out the suspect, who is always a local person.

A witch cannot operate at a distance as some form of contact is necessary for the evil to work. Suspects then are neighbours ('a witch is always closer than you think'). Because open accusations are never made, and because many people profess not to believe, and because those who do believe try to keep it to themselves, there is no common agreement that someone is a village witch. This contrasts, for example, with Franconian Switzerland.[22] Witchcraft is not a matter for the community, but for the two families, witch and bewitched, only.

Favret-Saada, whose fieldwork involved her in situations of 'witchcraft' diagnosis and cure, believes that few, if any, peasants practise witchcraft techniques. However, if someone is suspected by his neighbours he cannot fail to be aware of it. Peasants take an intense interest in the comings and goings of their neighbours. The innocent suspect's knowledge that someone he believes to possess supernatural power has been employed against him may well arouse anxieties that influence his behaviour, his health, and his enterprises. Rejecting the unwitcher's view of himself, he may come to blame his resulting problems on witchcraft in his turn.

Case X-6

Jean Babin was a married farmer, aged 38, an impotent alcoholic who suffered from depression. He could not consummate his marriage, although there was nothing wrong with his reproductive organs. He had a long series of misfortunes with his farm. Cattle died, cows aborted and suffered brucellosis, and crops suffered inexplicable blights. Jean married the sister of his brother's wife in 1963, when he was 33 years old. Three weeks before he married he was struck by a falling beam. He and his family blamed the accident, and the impotence that they believed was the result, on the witchcraft of Jean's neighbour, an old man, Ribault. The reason for the bewitchment was Jean's rejection of Ribault's suggestion that Jean marry his maidservant. The Babins claimed that Ribault had extraordinary sexual powers and seduced all his maidservants. He wanted Jean to marry this one so that she would be his neighbour and would always be available to him as a mistress. A witch's power makes him super-potent and avid for everything. The Babins claimed Ribault's wife was also a witch – 'she is getting it from her farmhands, sometimes five at one time!' Ribault was said to have responded to Jean's rejection with an ambiguous threat – 'In four or five years time, things will be sad.' Other comments attributed to Ribault were subsequently interpreted as threats or predictions of misfortune. It is possible that an unwitcher helped Jean Babin single him out from a list of possible witches.

22. See Sebald 1978: 103–8, the case of 'Mrs Meddler'.

The Babins claimed Ribault had two of the characteristics diagnostic of a witch. He seemed immune to death. They said he suffered illnesses that would kill an ordinary person. And his eyes conveyed omnipotence. (He had 'glassy eyes'.) Josephine Babin avoided him after they concluded that he was a witch, but her husband could not do so. Any time they had contact Jean believed that he later suffered some misfortune. Soon, any association with Ribault produced extreme nervousness in Jean (we are reminded of Sebastian and Stürmlin, in our initial case study), which led to extensive bouts of drinking.

However, Jean came to attribute some of the misfortunes that beset his farm to his uncle, Chicot. It was believed that he was displeased because Jean had taken over the farm. Jean found a priest who believed in witchcraft.[23] He blessed the farm and the misfortunes ceased, as did visits from Jean's uncle. To the Babins, the ultimate proof of Chicot's guilt was provided by the bizarre circumstances surrounding his death. He was sitting in a café when a complete stranger entered and denounced him and his companions and predicted that he would die within a week. He did. But despite the fact that several unwitchers had lifted spells off Jean and one witch (Chicot) had died, he remained impotent.

Marie of Izé, a noted unwitcher, refused to take on Jean's case, because she claimed that the spell was too strong for her, but she gave him a charm to help him. When Marie of Izé died of cancer, Jean believed she had been killed by a witch who proved too powerful for her, and that the charm would no longer work. He went on a drinking spree, and ended in a psychiatric hospital, where he came to the attention of Jeanne Favret-Saada, who was attempting a study of witchcraft in the Bocage. The Babins asked Favret-Saada to be their unwitcher, as they assumed that only someone who had 'power' would ask questions about witchcraft. She declined, but later introduced them to Madame Flora, an unwitcher with whom she was working. In the meantime Jean had managed to control his fits of nervousness by carrying charms of salt. He and his wife had both had encounters with Ribault and managed to face him down, and they believed he was leaving them alone as a consequence.

When she realized that Favret-Saada would not act as their unwitcher, Josephine Babin confided to her that before their marriage Jean had been falsely accused of being a witch by a neighbour in a private confrontation. The accusation was indirect and ambiguous, but to Jean its meaning was clear. Shortly afterwards he started drinking heavily and developed a bad case of eczema. The Babins believed that both were the consequence of a spell laid upon him by his neighbour's unwitcher, who had wrongly

---

23. Briggs points out the long historical association in France between local curés and anti-witchcraft, despite the Church's opposition (Briggs 1989: 22–4).

diagnosed him as the witch responsible for his client's misfortunes. Jean knew that he was innocent and possessed no magic book or charms for witchcraft. However, when Ribault bewitched him he felt unable to reveal the accusation to his various unwitchers, for fear of alienating them. Instead he assumed an air of disbelief about witchcraft that caused them to give up his case. Favret-Saada's diagnosis of Jean's impotence is that it was an unconscious reaction to his marriage and to his taking over his father's farm, both of which she believes had been forced upon him against his will (Favret-Saada 1980: 99–193).

–11–

Our Contemporary Witches

We have followed the story of the witch from small-scale communities to contemporary Europe. However, although traditional ideas of witchcraft have been relegated to the religious and geographical peripheries of industrial societies, witches have remained centre-stage. Indeed, they have often been given a greater importance than in the past. But as society and economy have become more complex and beliefs have become more secular we have changed their outward form.

Before discussing our contemporary 'witches', let us recapitulate what our case studies have shown us. The idea of witchcraft as a mystical ability practised by persons who often possess other non-natural attributes, such as the ability to fly or transport themselves in other unnatural ways, is related to the belief that the world is governed by personalized forces. In societies with such a world-view, witchcraft frequently is a way of explaining the problem of unmerited misfortune, and the image of the witch often epitomizes evil. Witchcraft accusations are directed at particular categories of persons.

It is rare for the witch-label to be attributed by the community-at-large to any particular accused, because there is often no consensus on an individual's guilt. In this situation, the emphasis in disputes involving witchcraft accusations usually is on attempting reconciliation, or upon severing the relationship involved. Only when there is widespread agreement that a person is a witch is the witch-image likely to be applied and punishment demanded. Where this happens, the accused is likely to occupy a vulnerable position with relation to many members of the community. Social groups or categories whose members lack full access to that which is valued in a society – such as the aged and the marginal – are particularly vulnerable to accusations. So too are those who have special magico-religious knowledge, or behave antisocially or in an unusual manner. A community-wide reputation builds up by a slow process of accusation and labelling, which often begins because of the behaviour of the suspect. However, under conditions of communal misfortune witch-hunts may take place, and it is possible that under these circumstances the process of labelling may be radically shortened and there may be less questioning of whether an accused is guilty. Because they are related to misfortune, anxiety, and conflict, witchcraft accusations are likely to be particularly common in periods of general social stress associated with communal calamities or unwelcome social change.

Witchcraft accusations may be used to discredit a rival in political competition. Because people believe that witchcraft exists and is motivated by malevolent feelings, the accusers may believe their accusation; but even tribal societies provide evidence that people recognize that accusations can be used cynically, as political strategies. When accusations are made in conditions of real power and wealth and involve persons of outstanding social status, the situation is a different one. The process of labelling is shortened considerably, and witches are held responsible for communal evils. The witch-image is applied to the accused, and the emphasis is on punishment rather than reconciliation. Among the Efik, for example, intense competition for great power and wealth involved political manipulation of witchcraft accusations in order to have one's rival accused and forced to submit to the poison ordeal. In all monarchies witchcraft against the king was a capital offence.

Witchcraft beliefs are at their most dangerous when they become linked to the pursuit of power and wealth. Where there exist marked differences in wealth and power between social classes, and powerful institutions and interest groups, witchcraft accusations operate politically in many ways. In late medieval Europe political competition between rival religious groups and the use of heresy charges for political purposes caused the witch-image to become elaborated, and this facilitated the European witch-crazes of the sixteenth and seventeenth centuries. From out of the political use of accusations, experts on witchcraft and witch-detection formulated an official ideology of witchcraft more complex than the popular ideology. In continental Europe the verdicts of these professional witch-finders no longer had to accord with community opinion, although it directed their search for suspects. As the society contained literate classes, the official ideology was codified and contained in works such as the *Malleus Maleficarum*. It focused on the witch-image and stressed punishment. Once codified, it influenced the nature of local accusations, so that even accusations between neighbours might result in trial and punishment.

Witch-Beliefs and Social Change

As witchcraft accusations are associated with tension and social conflict, and because social changes often generate conflicts and anxieties, we may expect periods of rapid and intensive change in small-scale societies to be characterized by heightened concern about witchcraft. Many of the cases given in Chapters 2 to 8 come from the early part of the colonial period, or, in the case of North America, shortly after Native American populations were subjected to Euro-American domination. This was a time of enforced acculturation and social change. Correspondingly, it was often a time of increased witchcraft accusation and witch-hunting, with new forms of tension and new patterns of accusations (Gluckman 1959: 101–2). Consequently many of our cases are from societies that may have been exhibiting higher incidences of witchcraft accusation than in precolonial times.

This is acknowledged in the studies on witchcraft among the Navaho (Kluckhohn 1967) and the Mohave (Devereux 1961), for example. Kluckhohn suggests that during their history of contact with United States institutions the Navaho have experienced two periods of major changes, both of which were characterized by an increased incidence of witchcraft accusations. These were in the 1870s, when the Navaho had been subjugated by the US army and were being brought under the effective control of the United States government, and in the 1930s and 1940s, when economic depression in the USA intensified the pressure of American society on the Navaho (Aberle 1966: 353; Kluckhohn 1967: 114–17). The former was the time of Manuelito's and Ganado Mucho's 'witch-hunt'. The latter was the period during which Kluckhohn was collecting and publishing his data.

In the early part of the colonial period new, imposed, alien governments applied legal sanctions in an attempt to prohibit witchcraft accusations, and ordeals were banned. These governments did not accept the existence of witchcraft, and measures taken against individuals believed to be witches became legal offences. Consequently, not only did local populations believe the numbers of witches had increased, but it was also now difficult and dangerous to take action against them, and this was regarded as one of the reasons for their increase. Under these circumstances new measures for coping with witchcraft developed. They included anti-witch movements, often embraced with enthusiasm. Some of these movements were initiated by traditional witch-finders. Others were initiated by new religious leaders, such as prophets, and were part of new religions (Mair 1969: 160–79).

In Chapter 4 I mentioned anti-witch movements among the Lele of Zaire, which took over some of the functions of the outlawed poison oracle (Douglas 1963). Another example is the *Atinga* cult that occurred among the Yoruba of south-western Nigeria in 1950–1 (Morton-Williams 1956). It originated in the Gold Coast (today's Ghana) in the 1940s, and spread into Nigeria from neighbouring Dahomey. Its leaders made and sold medicines, which when eaten gave the recipients protection from witchcraft as long as they observed four basic moral prohibitions. They should not steal, commit adultery, commit murder, or think evil against anyone (Morton-Williams 1956: 317). Should they attempt witchcraft or sorcery after taking the medicine, the *Atinga* spirit would kill them.

As well as offering protection against witchcraft, the movement identified witches. Youths and girls who were members of the cult became possessed by the *Atinga* spirit and identified certain women as witches. The majority of women accused were old. An accused who refused to confess was made to submit to the *Atinga* oracle. A chicken was sacrificed, and if it expired in any position other than upon its back ('facing heaven') the oracle had pronounced her guilty. She was expected to confess and produce her witchcraft paraphernalia. (Many of those accused produced everyday household ritual objects.) She was cleansed of her witchcraft by being washed from the *Atinga* medicine pot and eating the *Atinga* medicine.

Many old women confessed and were cleansed. Usually they claimed they must have become witches involuntarily and had been unaware of it. Once they accepted their guilt they tended to interpret specific events in their pasts – in particular, deaths of their children – as consequences of their witchcraft and to confess to these acts. Some women who refused to confess were tortured until they did so, and it was rumoured that some were beaten to death (Morton-Williams 1956: 315–26).

In precolonial times Yoruba witches always were women,[1] but in those days only a very small proportion of women were accused, and then only when some specific event precipitated an accusation. Normally, witchcraft was not dealt with through accusations against individuals. It was treated as a general phenomenon, and countered by community ancestor cults.[2] By the time of *Atinga* these cults were in decline, and were not accepted by Christian or Moslem Yoruba. But Christian, Moslem, and 'pagan' Yoruba alike were concerned about witchcraft and supported *Atinga*. *Atinga* identified as witches large numbers of old women, and spread rapidly throughout the Southwestern Yoruba.

Among the Shona, in addition to traditional diviners, prophets of the new Pentecostal Churches now divine causes of illness and identify witches. The Pentecostal sects believe in witchcraft, and offer hope to the accused by holding out the prospect of curing her through baptism, should she be contrite. Pentecostal prophets identify and cast out witch-spirits

---

1. Morton-Williams gives a detailed sociological analysis, with psychological undertones, of why Yoruba suspect elderly women of witchcraft (Morton-Williams 1956: 326–34).

2. I am informed by Professor Morton-Williams that it was common in African witchcraft for witchcraft as an ever-present menace to be countered by regular, institutionalized rites intended to reduce the powers of witches, or to placate them. Examples are the Nupe and the Yoruba. (Personal communication.)

possessing their followers. They may be consulted by non-Pentecostalists about the cause of illness, and when they are they seek to acquire as much knowledge as possible about their client's situation and personal relationships before giving a diagnosis, as does the traditional diviner (Crawford 1967: 221–43).

However, new witch-finding movements and institutions usually are not merely reactive. As well as being responses to new situations, often they are adaptive. They may enable persons to adapt to changed social circumstances, and are themselves a part of the process of social change. Keith Thomas and Alan Macfarlane have made this point with respect to the operation of witch beliefs in sixteenth- and seventeenth-century England (Macfarlane 1970a; Thomas 1973).

Early functionalist explanations of witchcraft beliefs and accusations envisaged them as operating within a context of stability, even where they were seen as promoting changes in relations between persons and between groups.[3] More recent analyses often see witchcraft as operating within the context of radical social change, and being not simply an *expression of* change, but a *contributor to* change. The new anti-witchcraft movements become modified as they develop, and may help to alter existing beliefs. As a consequence, the importance given to witchcraft may come to be reduced, or old patterns of belief and accusation may be replaced by new ones.

In traditional Navaho society witchcraft beliefs operated as a significant social leveller. The rich treated the poor generously out of fear of their witchcraft, and in general acted generously for fear of being accused themselves. In the late 1930s economic changes began to erode the traditional network of reciprocal relationships between individuals and groups, which was supported by aspects of Navaho witchcraft beliefs. Today the family has become more independent, and many people desire to be free of traditional demands. The peyote religion, the Native American Church, has been widely adopted, and it attacks traditional relationships and associated witchcraft beliefs in a number of ways.[4] Peyote priests may diagnose witchcraft. Sometimes they name ceremonial singers as witches. Peyote attacks traditional ceremonial ideas. Because of this there is antipathy between singers and peyote priests. Peyote is believed by many of its followers to cure the effects of witchcraft attacks, and some peyotists claim that peyote will cure a witch of his witchcraft. Others claim peyote has told them that witchcraft is an illusion. Peyote also provides security against the dead. In all these ways peyotism attacks

---

3. See Max Marwick's hypotheses in Chapter 6.
4. I do not wish to imply that this is the only, or even the main, reason for its adoption. I wish simply to draw attention to some of the consequences.

the power of witchcraft beliefs and other traditional aspects of Navaho society, and helps many Navaho to adapt to a changing social situation.

Large numbers of Navaho have become peyotists. (In 1964, estimates by peyotists varied from 25 to 50 per cent of the reservation population.) Navaho in the northern part of their reservation have a history of greater opposition to Whites and to the Indian Service, and are more committed to traditional religion. Those in the southern part exhibit a greater acceptance of the social changes, and have adopted peyote more readily (Aberle 1966: 179, 184–6, 193, 203–4, 218–19, 353).

Significantly, when the witch-finding activities of *Atinga* were suppressed by the colonial authorities, the movement turned its attention to attacking certain types of local gods (*orisha*), and destroyed their symbols, images, and shrines. It attacked those gods that represented the relationship between Man and nature. (It did not attack the ancestor cults, and in this it showed respect for the values of lineage and community.) The Yoruba were no longer so subject to the vagaries of nature, but they still feared these *orisha*, which were under the control of the more conservative members of the oldest generations. The attack upon them can be seen as an assertion by youth of the changing cultural outlook, and its rejection of conservative age. Members of the older generations who sponsored and supported the *Atinga* cult had benefited from the political and economic changes in Yoruba society, and stood to gain if the heads of the traditional *orisha* cults lost their influence. Although it began as a witch-hunting organization, in its last phase *Atinga* became an assertion of the reality of social and cultural changes taking place in Yoruba society (Morton-Williams 1956: 325–6, 332–4).

If increased fear of witchcraft is associated with social change, we might expect to find it in the developing towns of colonies and colonial societies in a large measure. Again, however, we find witchcraft ideas and accusations being modified by new social circumstances, and often giving way to other mystical explanations for misfortune. In the 1940s, 1950s and 1960s, migrants into the towns of Southern Africa continued to interpret misfortune in personal terms, looking for causes within their personal relationships. Initially, they frequently accepted explanations in terms of witchcraft. However, the social structure of townships and the opposition of European authorities to witchcraft beliefs often prevented victims from obtaining satisfaction for misfortunes so diagnosed. Clyde Mitchell, analysing explanations of misfortune used by urban Africans in Southern Africa, found that over time there was a development away from explanations by witchcraft towards other mystical causes. In the cases he studied the development was towards explanations in terms of ancestral spirits. Sufferers believed that this enabled them to take effective remedial action. Witchcraft remained a common explanation only between persons who saw themselves as being in direct competition for social or economic ends (Mitchell 1965).

In two recent articles Rowlands, Warnier and Fisiy examine the changing role of witchcraft and witchcraft accusations in the contemporary state of Cameroon. Traditionally, fear of accusations was an important mechanism that enforced redistribution of goods and services, as wealth and other forms of success demonstrated possession of occult power. Today, members of political élites are pressured to assist their home villages through fear of being attacked by witchcraft should they fail to do so. At the same time witchcraft has become recognized as a crime by the State, and convicted persons are liable to legal punishment. Rowlands and Warnier interpret these contrasting developments as expressions of the tension between the peasantry and the new economic and political élites. In Developing Countries, witchcraft beliefs and patterns of accusations have been modified in the process of social and economic change. They have not necessarily decreased in social importance (Fisiy and Rowlands 1989: Rowlands and Warnier 1988).

The 'Witches' of Contemporary Industrial Society

In contemporary industrial societies, in general, we no longer believe in the traditional idea of the witch. We no longer believe in the existence of persons who practise evil by supernatural means, change into animals, control familiars, or ride the wind. But nonetheless we often believe in the existence of persons who set out deliberately to cause harm, out of malice or as a result of their very natures. Often they are believed to exhibit many of the other characteristics of the witch, such as behaving in ways that invert or caricature normal, proper behaviour, seeking deliberately to undermine morality, causing personal misfortunes, and conspiring to destroy us or our world.

These persons are the scapegoats of our society. Again, their selection is not random. They belong to the same social categories from which are selected the witches of small-scale communities and of colonial and medieval and Early Modern times and their selection is based on the same logic. For some reason it is felt that they may hold a grudge against society – because they are deprived, or do not 'belong', or behave in ways that many consider outrageous or antisocial. The way they are viewed and the ways they are believed to behave and to seek to achieve their aims are all 'mystical' in the sense that they defy routine, everyday experience. It is this that places relations with such persons in a different category from 'normal', 'ordinary' relations, and enables us to say that they are the 'witches' of industrial society.

It appears that 'witchcraft' accusations are less common in

interpersonal relationships in industrial societies than they are in many tribal societies and closed communities. This is for the same reason that they are rare among pastoralists (Baxter 1972), which was discussed in Chapter 2. The operation of a variety of processes means that persons who come to dislike one another often do not have to interact. Even neighbours may have little contact with one another. Even so, we should not minimize the degree to which 'witchcraft' ideas may become involved in interpersonal relations. We are all familiar with the hatreds and suspicions that can develop between close relatives in our society, and many of us are probably aware of hostile neighbours who have developed fantasies of witchcraft proportions about each other.

As in other cultures and societies, where there is widespread agreement that specific individuals are 'witches' it is because they belong to particular categories of persons. They may be political enemies or belong to antisocial categories, but in 'modern' society the persons most prominently treated in this way are members of racial or ethnic minorities. They hold a common relationship with many members of the wider society. They are the strangers within society, or, increasingly, the groups that, whilst living within society are perceived as being not properly *of* it. I believe the crucial point about such minority groups is that they are structurally integrated into society in an ambiguous or/and a marginal position. This makes their position different from that of the purely external 'enemy', with whose nationality they may be identified. Although the members of a society may regard nations living beyond their borders as hostile, they rarely regard them as 'evil', whereas they often perceive their own racial and ethnic minorities in this way. Whilst there may be an element of xenophobia in this perception, it is a situation that goes beyond xenophobia.

Racial and ethnic minorities often are excluded from full membership of a society, and this is seen by the public as a further reason why they would hold attitudes and aims inimical to the welfare of the society. Frequently, the public rationalizes their exclusion as being their own fault. They may become the focus of witchcraft beliefs and accusations, particularly in periods of social and economic change and uncertainty, when many people may come to hold them in large measure deliberately responsible for personal and social problems.

Today in industrial societies, many of those misfortunes that in small-scale societies and communities, and in developing societies, may be explained as the consequence of witchcraft, such as illness and accident, are recognized to result from other circumstances. But there are new misfortunes that, whilst affecting large numbers of the population, nonetheless involve a personal or chance element. They include unemployment, reduction in the value of earnings and property, and decline in living standards. And there are other, less tangible problems, such as loss of national prestige, that are more difficult for people to assess but which they may feel keenly. Even where individuals recognize an impersonal scientific explanation for a misfortune such as an illness, its transmission may be blamed upon others. Members of minority groups may be blamed for misfortunes that can be interpreted as affecting oneself in particular. They have the job that should rightfully be yours, or they have caused the value of your house to fall, or their dirty and unsocial and abnormal behaviour spreads disease that could be caught by yourself. They are being blamed for problems that are the consequences of social processes such as the vagaries of the labour market or the housing market. Accusers may not understand this. Or they may find it more satisfying to blame individuals and groups rather than abstract social principles or the relatively impersonal actions of institutions or faceless persons, against which they are powerless.

As minorities are perceived as relatively undifferentiated groups, witch-beliefs concerning members of a minority stigmatize the whole group, and the process of labelling is a shortened one. When such beliefs are used in situations of political and economic competition or are harnessed by political groups and leaders, minorities are presented as responsible for problems regarded as affecting the whole population, such as the general level of unemployment or national political decline. The emphasis now is on the communal consequences of the 'witches'' actions, and not just the personal ones. Under these circumstances a real anti-'witch' solution may be propounded. There will be demands for restrictions on the witch-group, and there may be calls for their expulsion from the society, or even their extermination.

Such situations are commonplace in recent European history. With the current break-up of the Soviet empire they are likely to become more overt in eastern Europe, with the open expression of anti-Semitism and the persecution of ethnic minorities. The most outstanding example from recent history is the Nazi persecution of Europe's Jews (Cohn 1967).

Between the second and fourth centuries, the Christian Church created the fantasy of the Jews as a brotherhood of evil in order to strengthen its position in the competition between Church and Synagogue for converts in the Hellenistic world. Some eight centuries later the Roman Catholic Church developed these fantasies into a demonology, claiming the Jews were employed by the Devil to undermine Christianity and collectively rewarded with mastery of black magic. From the twelfth century Jews were perceived as a conspiracy of sorcerers that possessed limitless powers of evil and was seeking to destroy Christendom on Satan's orders. When the Antichrist came he would be a Jew (Cohn 1967: 20–3, 253–4).

In medieval times, before the idea of the witch became associated with apostasy, so that the witch had to be a Christian who had renounced Christ, Jews were often used as witch-figures. For example, they were blamed for introducing the plague into Europe in the middle of the fourteenth century (Ginzburg 1992: 33–86). At a Synod in 1317, Bishop de Ledrede, whose pursuit of Lady Alice Kyteler may have contributed to the developing idea of the apostate witch, inveighed against 'a certain new and pestilential sect in our parts, differing from all the faithful in the world, filled with a devilish spirit, *more inhuman than heathens or Jews*, who pursue the priests and bishops of the Most High God equally in life and death, by spoiling and rending the patrimony of Christ in the diocese of Ossory' (Seymour 1992: 47).

In the most virulent form of modern anti-Semitism, the idea of Jews as a brotherhood of evil persists in the belief in an organized Jewish conspiracy to destroy the civilized world and replace it with a world order dominated by Jewry (Cohn 1967: 61–5).

> The myth of the Jewish world-conspiracy represents a modern adaptation of this ancient demonological tradition. According to this myth there exists a secret Jewish conspiracy which, through a world-wide network of camouflaged agencies and organizations, controls political parties and governments, the press and public opinion, banks and economic developments. The secret government is supposed to be doing this in pursuance of an age-old plan and with the single aim of achieving Jewish domination over the entire world; and it is also supposed to be perilously near to achieving this aim (Cohn 1967: 22–3).

The European conception of the witches' sabbat is replaced by a more 'secular' idea of a secret Jewish government working to dominate the world. To the Nazi ideologists, this Jewish world conspiracy was the consequence of an innate quality of Jewishness, an inherent will to evil present in every Jew. History was a continuing battle between the force for evil, embodied in the Jew, who is aided by his auxiliaries – Bolshevism, Freemasonry, and sections of the Church, and the force for good, embodied in the Aryan or Nordic race (Cohn 1967: 178–80). Evil could only be extirpated by exterminating the Jews, because their evil was an innate quality.

In *Mein Kampf*, Hitler writes of the Jews in the language of pestilence – of an evil that cannot be compromised with and which must be extirpated.

> [The] Jew does not ever think of leaving a territory which he has once occupied. He sticks where he is with such tenacity that he can hardly be driven out, even by superior physical force.... He is and remains a parasite, a

sponger who, like a pernicious bacillus, spreads over wider and wider areas according as some favourable area attracts him. The effect produced by his presence is also like that of the vampire; for wherever he establishes himself the people who grant him hospitality are bound to be bled to death sooner or later (Hitler 1939: 255).[5]

But a Jew can never be rescued from his fixed notions (Hitler 1939: 63).

There is no such thing as coming to an understanding with the Jews. It must be the hard-and-fast 'Either–Or' (Hitler 1939: 178).

And so I believe today that my conduct is in accordance with the will of the Almighty Creator. In standing guard against the Jew I am defending the handiwork of the Lord (Hitler 1939: 179).

This is the real language of witchcraft. It is the language of the accuser, not of the accused. The witch's crime is truly 'a deed without a name'.

Without the social and economic effects of the world depression Hitler and the Nazis never would have come to power (Cohn 1967: 198). Once in power Hitler was able to propagate his ideas further and put them into effect. The great majority of the German people never were obsessed with the idea of a Jewish world conspiracy (Cohn 1967: 212), but the power wielded by the Nazi government enabled the fanatics and their henchmen to seek to extirpate the 'witch' at the centre of their dualistic conception of the world.

In contemporary Britain, immigrant communities from the New Commonwealth – principally from the West Indies, India, Pakistan, Bangladesh, and East Africa – commonly are portrayed, in the media and in the political discourse of the New Right, as a threat to Britain's 'national character'. It is claimed that they do not share the social and moral values of 'the British nation', and that this is a cause of racial tension, and therefore they are to blame for creating certain problems and tensions within the society. A powerful strand in the language of political and popular discourse on racial matters in Britain presents the 'black community' as an 'enemy within' that attacks the society from its position inside it. This is the classic language of witchcraft.

In recent years an emphasis in right-wing ideology in Britain has developed which argues that antipathy towards persons of different race or culture is 'natural', and that consequently attempts to build a multiracial society are doomed to create conflicts which threaten overall stability. In this way, the image of the 'black' person comes more to resemble the

---

5. This is by no means the most virulent anti-witch language used in *Mein Kampf* against the Jews. See, for example, Hitler 1939: 273.

classic image of the witch. He carries 'evil' as a part of himself (Solomos 1991). However, in Britain the witch-image of the 'coloured immigrant' has not achieved its potential. (In parts of the European continent, on the other hand, the process appears to have gone further than in Britain, with a greater antipathy towards foreign immigrants and workers.) Currently, racial and ethnic minorities are not blamed to any significant degree for economic recession, unemployment, and other problems adversely affecting large numbers of citizens, and there appears to have been a decline in the importance of racial issues in British politics in the late 1970s and the 1980s.

Some analysts argue that this is because recent governments have presented the trade unions as the main threat to the stability of British society, and this has re-established class relations as the main arena of social conflict (Rex 1987: 363). Significantly, the language used to stigmatize the trade unions has also been that of the 'enemy within'. However, neither ethnic minorities nor trade unionists have become full-blown 'witches'. I suggest that any widespread scapegoating has the potential to develop into a witch-hunt in which the typical witch-image is applied to the accused. Then the accused group is believed to seek actively to subvert the moral foundation and political order of society, and the behaviour of its members is held to be an inversion or caricature of 'proper', 'decent' behaviour. Whether or not this potential is achieved depends upon the social circumstances within which the scapegoating is operating. For example, whether there is widespread and increasing social and economic deprivation and, in particular, the manner in which the scapegoating is used politically.

In a recent article, Professor Mary Douglas has compared the use of witchcraft accusations to accusations of leprosy in North Europe in the twelfth century. She suggests that a leprosy epidemic at this time did not exist – its appearance is an illusion created by the use of leprosy accusations by the community in order to stigmatize and remove landless persons and heretics from society (Douglas 1991). Reference to leprosy reminds us of AIDS. Groups regarded as being at particular risk of contracting AIDS are in danger of becoming the 'lepers' of contemporary society. Homosexual men especially have been the target of this kind of attitude. In the United States, where AIDS first became popularly identified as a disease of gay men, right-wing moralists such as Moral Majority groups and fundamentalist Christian organizations have presented AIDS as a type of just retribution for a morally degenerate lifestyle. (In England, a police Chief Constable has characterized AIDS sufferers as 'swimming in a sewer of their own making'.)

However, gays are blamed not only for creating AIDS through unnatural behaviour. Critics of their lifestyle also claim they may be transmitting the virus to the heterosexual population through irresponsible behaviour. They are not only accused of having produced the disease, they are said to be spreading it to innocent victims in the non-homosexual population by behaviour that is at the least uncaring, and at the worst deliberate. Thus not only do some religious groups see gays as constituting a morally inverted community, they portray them as deliberate purveyors of evil (Altman 1988: 1–81; Sabatier 1988: 90–5).

We may expect the AIDS epidemic to increase 'witchcraft' accusations against non-conforming groups such as homosexuals, and the United States and other industrial countries present evidence of increasing discrimination against gays and attacks upon them, related to fear of AIDS (Altman 1988: 24–65). Moreover, in the United States AIDS is becoming disproportionately a disease of the African-American and Latino communities, because it flourishes under conditions of social deprivation. There is now a serious risk that White Americans might come to see African-Americans and Latinos as AIDS carriers, with all the potential for witch-labelling (Sabatier 1988: 122–37). The Haitian minority in the United States has already suffered stigmatization and discrimination through being incorrectly identified as a 'high-risk group' for AIDS (Altman 1988: 36–9, 71–4; Sabatier 1988: 86), and racist groups in the United States have attacked homosexuals and non-White minorities as purveyors of AIDS into the White community through 'unnatural' sexual practices – under which description they include interracial heterosexual relations (Sabatier 1988: 90–5).

It is not true that industrial man is too sophisticated to believe in witches, but merely that many of us are too sophisticated to believe in witches that fly.

In my examination I have taken the view that the witch, as an individual effectively propagating misfortune by mystical means, does not exist. And yet clearly there is a sense in which he or she does exist. Our cases involved real people who were identified with this mystical offence. From Stürmlin to the Jews in the Nazi extermination camps, their offence was believed to be real enough, and consequently so were the punishments they were made to suffer. Just because the witch is an invention, we cannot dismiss the witch-belief simply as a delusion or a fantasy. We cannot do so because the witch is a *social* invention. He or she is a product of the nature of social relationships, of the fact that people desire, and see themselves as competing for, valued and finite goals, such as health, youth, wealth and power. Consequently, in most societies some individuals accuse other individuals of witchcraft, in a continuously recurring process. Whether or not other persons believe their accusations is influenced by cultural beliefs and social factors. And sometimes we have witch-hunts, where the leaders or representatives of social groups seek out witches in their name. This is a process that can explode again and again in human society. Although we may like to think of it as something in our past history, when we examine the witch-belief in the context of social scapegoating we realize that it is a phenomenon that has no historical ending and that may flare up into major hunts at almost any time, but particularly during periods of social stress and social tension. If we know more about the witch-phenomenon and if we know what categories of people are likely to be accused of this mystical offence, then we may be able to arm ourselves to prevent the application of the witch-label. But we can only be reasonably sure of success if we can prevent the political, social and economic conditions that encourage its widespread application and sometimes place the witch-finders in political control.

Glossary

Adelphic succession. A man is succeeded in office by his younger brother.

Affinal relative. A relative by marriage.

Agnatic descent. Genealogical descent through males only. (Also termed *patrilineal descent*.)

Agnatic tie. The kinship relationship between two persons (male and/or female) who are related through intermediary *male* relatives – for example, Father's Brother's Son/Father's Brother's Daughter.

Clan. A socially recognized kinship group to which membership is obtained ideally *either* through one's father or one's mother, but all of whose members are not able to trace their genealogical relationship with each other (in contrast to a *lineage*).

Classificatory brother. A relative whom one classifies in the same category of relative as one's brother, and calls by the same term.

Classificatory mother's brother. A relative whom one classifies in the same category of relative as one's mother's brother, and calls by the same term.

Exogamous clan. A clan whose members are not allowed to intermarry but must take their spouses from outside their clan.

Functionalism refers to a set of theories whose basic premiss is that an institution exists because it fulfils some basic societal need. (See, for example, Mair 1969: 199–203.)

Institutions. Widely held beliefs and standardized patterns of behaviour in a culture, such as witchcraft and witchcraft accusations.

Interests. Those things that people regard as desirable. They can be material or non-material. They are a consequence of man's membership of society and of groups within it (Marshall 1994: 411–13).

Lineage. A socially recognized kinship group to which membership is ideally obtained *either* through one's father or one's mother, all of whose members are able to trace their genealogical relationship with each other (in contrast to a *clan*).

Malefici. Practitioners of *maleficium*.

Maleficium. Causing harm by occult means. (plural, *maleficia*.)

Matriclan. A socially recognized kinship group that consists ideally of the descendants, male and female, in the female line only from a

Glossary

common ancestress, not all of whose members can trace their genealogical relationship to one another (cf. *matrilineage*).

Matrilineage. A socially recognized kinship group consisting ideally of all the descendants, male and female, in the female line only from a common ancestress, all of whom can trace their genealogical relationship with each other.

Matrikin. Any relatives on one's mother's side. (Where a society contains matrilineal groupings, then matrikin will include one's matrilineal kin – the members of one's *matrilineage* or *matriclan*.)

Matrilineal. Descending through the female line.

Patriclan. A socially recognized kinship group that consists ideally of the descendants, male and female, in the male line only from a common ancestor, not all of whose members can trace their genealogical relationship to one another (cf. *patrilineage*).

Patrilineage. A socially recognized kinship group that consists ideally of the descendants, male and female, in the male line only from a common ancestor, all of whom are able to trace their genealogical relationship with each other.

Patrilineal clan. See *patriclan*.

Phratry. Where a society contains a number of clans, and these are grouped into even larger descent-based units, then where there are *three or more* such larger units each of these is termed (by anthropologists) a *phratry*.

Polygyny. The socially accepted practice of a man being allowed more than one wife at the same time.

'"*Power*" is the probability that one actor within a social relationship will be in a position to carry out his own will despite resistance, regardless of the basis on which this probability rests' (Weber 1966: 152).

Shaman. A curer who operates by being possessed by spirits or by a mystical power.

Small-scale societies or communities have relatively little differentiation within the community. They have a relatively simple technology, and knowledge is usually stored and communicated orally. Any member is in face-to-face relationships with many other members, and consequently relationships are strongly personalized. Individuals often occupy several different kinds of social positions *vis-à-vis* each other, and the predominant relations are usually those of kinship. Such societies or communities usually have small populations.

Tribe. That type of small-scale society comprising a number of autonomous political units sharing common linguistic and cultural features.

Virilocal residence. On marriage a woman goes to live with or near her husband's parents.

Glossary

Witch-crazes. Local outbreaks of witch-hunting that give the impression that the community has become obsessed with fear of witches. (As the outbreaks usually occurred several centuries ago, often it is not possible to know to what degree people had really become obsessed about witchcraft, or whether the impression is simply the consequence of the kinds of data historians have at their disposal.)

Bibliography

Aberle, D. F. 1961. 'Navaho', in D. M. Schneider and K. Gough (eds), *Matrilineal Kinship*, Berkeley and Los Angeles: University of California Press, pp. 96–201

Aberle, D. F. (with field assistance by Harvey C Moore). 1966 *The Peyote Religion among the Navaho*, 2nd edn, Chicago and London: University of Chicago Press

Alexander, J. C. 1987. *Sociological Theory since 1945*, London: Hutchinson

Altman, D. 1988. *AIDS and the New Puritanism*, London: Pluto Press

Anglo, S. 1977a. 'Evident Authority and Authoritative Evidence: the *Malleus Maleficarum*', in S. Anglo (ed.), *The Damned Art: Essays in the Literature of Witchcraft*, London: Routledge and Kegan Paul, pp. 1–31

Anglo, S. 1977b. 'Reginald Scot's *Discoverie of Witchcraft*: Scepticism and Sadduceeism', in S. Anglo (ed.), *The Damned Art: Essays in the Literature of Witchcraft*, London: Routledge and Kegan Paul, pp. 106–39

Apuleius. 1950. *The Golden Ass*, trans. by Robert Graves, London: Penguin Books

Barber, M. 1973. 'Propaganda in the Middle Ages: Charges against the Knights Templar', *Nottingham Journal of Medieval Studies*, pp. 42–57

Baroja, J. C. 1965. *The World of the Witches*, trans. O. N. V. Glendinning, Chicago: University of Chicago Press

Basso, K. H. 1969. *Western Apache Witchcraft*, Anthropological Papers of the University of Arizona, No. 15, Tucson: University of Arizona Press

Basso, K. H. 1979. *Portraits of 'the Whiteman'. Linguistic Play and Cultural Symbols among the Western Apache*, Cambridge: Cambridge University Press

Baxter, P. T. W. 1972. 'Absence Makes the Heart Grow Fonder. Some Suggestions Why Witchcraft Accusations are Rare among East African Pastoralists', in M. Gluckman (ed.), *The Allocation of Responsibility*, Manchester: Manchester University Press, pp. 163–91

Bednarski, J. 1970. 'The Salem Witch-Scare Viewed Sociologically', in

M. Marwick (ed.), *Witchcraft and Sorcery*, London: Penguin Books, pp. 151–63

Beidelman, T. O. 1980. 'The Moral Imagination of the Kagaru: Some Thoughts on Tricksters, Translation and Comparative Analysis', *American Ethnologist*, Vol. 7, pp. 27–42

Beidelman, T. O. 1986. *Moral Imagination in Kagaru Modes of Thought*, Bloomington: Indiana University Press

Bohannan, L. 1958. 'Political Aspects of Tiv Social Organization', in J. Middleton and D. Tait (eds), *Tribes Without Rulers*, London: Routledge and Kegan Paul, pp. 33–66

Bohannan, L. and Bohannan, P. 1953. *The Tiv of Central Nigeria*, Ethnographic Survey of Africa, Western Africa, Part VIII, London: International African Institute

Bohannan, P. 1958. *Extra-processual Events in Tiv Political Institutions*, Bobbs-Merrill Reprint Series in the Social Sciences, No. A-17, Indianapolis: the Bobbs-Merrill Company Inc.

Boyer, P. and Nissenbaum, S. 1974. *Salem Possessed: the Social Origins of Witchcraft*, Cambridge Mass.: Harvard University Press

Briggs, R. 1989. *Communities of Belief: Cultural and Social Tension in Early Modern France*, Oxford: Clarendon Press

Bryant, A. T. 1929. *Olden Times in Zululand and Natal*, London, New York, Toronto: Longman Green

Burridge, K. M. 1965. 'Tangu, Northern Madang District', in P. Lawrence and M. J. Meggitt (eds), *Gods, Ghosts and Men in Melanesia*, Melbourne, London, New York: Oxford University Press, pp. 224–49

Canfield, G. W. 1983. *Sarah Winnemucca of the Northern Paiutes*, Norman and London: University of Oklahoma Press

Cardozo, A. R. 1970. 'A Modern American Witch-Craze', in M. Marwick (ed.), *Witchcraft and Sorcery*, London: Penguin Books, pp. 369–77

Carr, E. H. 1990. *What is History?*, London: Penguin Books

Clark, G. 1966. *Early Modern Europe. From About 1450 to About 1720*, London, Oxford, New York: Oxford University Press

Clark, S. 1977. 'King James's *Daemonologie*: Witchcraft and Kingship', in S. Anglo (ed.), *The Damned Art: Essays in the Literature of Witchcraft*, London: Routledge and Kegan Paul, pp. 156–81

Cohen, P. S. 1968. *Modern Social Theory*, London: Heinemann

Cohn, N. 1967. *Warrant for Genocide*, London: Eyre and Spottiswoode

Cohn, N. 1970. 'The Myth of Satan and his Human Servants', in M. Douglas (ed.), *Witchcraft Confessions and Accusations*, London: Tavistock, pp. 3–16

Cohn, N. 1976. *Europe's Inner Demons*, London: Paladin

Collins, R. 1994. *Four Sociological Traditions*, Oxford: Oxford University Press

Costain, T. B. 1973. *The Three Edwards*, London: Tandem Books
Craib, I. 1984. *Modern Social Theory. From Parsons to Habermas*, New York: Harvester Wheatsheaf
Crawford, J. R. 1967. *Witchcraft and Sorcery in Rhodesia*, International African Institute: Oxford University Press
Crick, M. 1973. 'Two Styles in the Study of Witchcraft', *Journal of the Anthropological Society of Oxford*, Vol. IV, pp. 17–31
Crick, M. 1976. *Explorations in Language and Meaning*, London: Malaby Press
Davies, R. T. 1947. *Four Centuries of Witch Beliefs. With Special Reference to the Great Rebellion*, London: Methuen & Co
Devereux, G. 1961. *Mohave Ethnopsychiatry and Suicide: the Psychiatric Knowledge and the Psychiatric Disturbances of an Indian Tribe*, Bureau of American Ethnology, Bulletin 175, Washington: The Bureau
Devons, E. and Gluckman, M. 1964. 'Conclusion: Modes and Consequences of Limiting a Field of Study', in M. Gluckman (ed.), *Closed Systems and Open Minds. The Limits of Naivety in Social Anthropology*, Edinburgh and London: Oliver and Boyd, pp. 158–261
Dolgin, J. L., Kemnitzer, D. S. and Schneider, D. M. (eds). 1977. *Symbolic Anthropology. A Reader in the Study of Symbols and Meanings*, New York: Columbia University Press
Douglas, M. 1963. 'Techniques of Sorcery Control in Central Africa', in J. Middleton and E. H. Winter (eds), *Witchcraft and Sorcery in East Africa*, London: Routledge and Kegan Paul, pp. 123–41
Douglas, M. 1966. *Purity and Danger. An Analysis of Concepts of Pollution and Taboo*, London: Routledge and Kegan Paul
Douglas, M. 1967. 'Witch Beliefs in Central Africa', *Africa*, Vol. xxxvii, No. 1, January 1967, pp. 72–80
Douglas, M. 1970. 'Thirty Years After *Witchcraft, Oracles and Magic*', in M. Douglas (ed.), *Witchcraft Confessions and Accusations*, London: Tavistock, pp. xiii–xxxviii
Douglas, M. 1973. *Natural Symbols. Explorations in Cosmology*, New York: Vintage Books
Douglas, M. 1991. 'Witchcraft and Leprosy: Two Strategies of Exclusion', *Man (NS)*, Vol. 26, pp. 723–36
Duerr, H. P. 1987. *Dreamtime. Concerning the Boundary between Wilderness and Civilisation*, trans. Felicitas Goodman. Oxford: Basil Blackwell
Dutton, B. P. 1975. *Navahos and Apaches: the Athabascan Peoples*, Englewood Cliffs: Prentice-Hall Inc.
Edmunds, R. D. 1985. *The Shawnee Prophet*, Lincoln and London: University of Nebraska Press
Evans-Pritchard, E. E. 1971. *The Azande. History and Political*

Institutions, Oxford: Clarendon Press

Evans-Pritchard, E. E. 1976. *Witchcraft, Oracles and Magic among the Azande*, (abridged with an introduction by Eva Gillies), Oxford: Clarendon Press

Favret-Saada, J. 1980. *Deadly Words. Witchcraft in the Bocage*, trans. Catherine Cullen, Cambridge: Cambridge University Press

Fisiy, C. F. and Rowlands, M. 1989. 'Sorcery and Law in Modern Cameroon', *Culture and History*, Vol. 5/6, pp. 63–84

Forbes, T. R. 1966. *The Midwife and the Witch*, New Haven and London: Yale University Press

Forde, D. (ed.) 1956. *Efik Traders of Old Calabar*, London: Oxford University Press

Forde, D. 1964. *Yakö Studies*, Oxford: International African Institute

Forde, D. and Jones, G. I. 1950. *The Ibo and Ibibio-speaking Peoples of South-eastern Nigeria*, Ethnographic Survey of Africa, Western Africa, Part III, London: International African Institute

Forge, A. 1970. 'Prestige, Influence and Sorcery: a New Guinea Example', in M. Douglas (ed.), *Witchcraft Confessions and Accusations*, London: Tavistock, pp. 257–75

Fowler, C. S. and Liljeblad, S. 1986. 'Northern Paiute', in W. L. D'Azevedo (ed.), *Handbook of North American Indians, Vol. 11: Great Basin*, Washington: Smithsonian Institution, pp. 435–65

Geertz, H. 1975. 'An Anthropology of Religion and Magic, I', *Journal of Interdisciplinary History*, Vol. 6, pp. 71–88

Gelfand, M. 1967. *The African Witch*, Edinburgh and London: E. S. Livingstone Ltd

Gillies, E. 1976. 'Introduction' to E. E. Evans-Pritchard, *Witchcraft, Oracles and Magic among the Azande* (abridged with an introduction by Eva Gillies), Oxford: Clarendon Press, pp. vii–xxxiii

Ginzburg, C. 1966. *The Night Battles. Witchcraft and Agrarian Cults in the Sixteenth and Seventeenth Centuries*, trans. John and Anne Tedeschi, London, Melbourne, and Henley: Routledge & Kegan Paul

Ginzburg, C. 1990. 'Deciphering the Sabbath', in B. Ankarloo and G. Henningsen (eds), *Early Modern European Witchcraft: Centres and Peripheries*, Oxford: Clarendon Press, pp. 121–37

Ginzburg, C. 1992. *Ecstasies. Deciphering the Witches' Sabbath*, London: Penguin Books

Gluckman, M. 1959. *Custom and Conflict in Africa*, Oxford: Basil Blackwell

Gluckman, M. 1965. *Politics, Law and Ritual in Tribal Society*, Oxford: Basil Blackwell

Gluckman, M. 1968. 'The Utility of the Equilibrium Model in the Study of Social Change', *American Anthropologist*, Vol. 70, No. 2, pp. 219–37

Gorman, F. J. E. 1981. 'The Persistent Identity of the Mohave Indians 1859–1965', in G. P. Castile and G. Kushner, (eds), *Persistent Peoples. Cultural Enclaves in Perspective*, Tucson: University of Arizona Press, pp. 43–68

Haggard, H. Rider. 1963. *Nada the Lily*, London: Macdonald

Hansen, C. 1971. *Witchcraft at Salem*, London: Hutchinson and Co.

Harley, D. 1990. 'Historians as Demonologists: the Myth of the Midwife Witch', *Social History of Medicine*, Vol. III, pp. 1–26

Harris, M. 1977. *Cows, Pigs, Wars and Witches. The Riddles of Culture*, London: Fontana Books

Hart, N. 1985. *The Sociology of Health and Medicine*, Ormskirk: Causeway Press

Harwood, A. 1970. *Witchcraft, Sorcery, and Social Categories among the Safwa*, London: Oxford University Press

Hitler, A. 1939. *Mein Kampf*, London: Hurst and Blackett Ltd

Holmes, C. 1993. 'Women, Witnesses and Witches', *Past and Present*, No. 140, pp. 45–78

Horton, R. 1967. 'African Traditional Thought and Western Science', *Africa*, Vol. 37, pp. 50–71, 155–87

Hough, P. 1991. *Witchcraft. A Strange Conflict*, Cambridge: The Lutterworth Press

Huxley, A. 1971. *The Devils of Loudun*, London: Penguin Books

Jennings, F. 1984. *The Ambiguous Iroquois Empire*, New York, London: Norton

Jones, G. I. 1956. 'The Political Organization of Old Calabar', in D. Forde (ed.), *Efik Traders of Old Calabar*, London: Oxford University Press, pp. 116–60

Jones, G. I. 1970. 'A Boundary to Accusations', in M. Douglas (ed.), *Witchcraft Confessions and Accusations*, London: Tavistock, pp. 321–32

Kieckhefer, R. 1976. *European Witch Trials: Their Foundations in Popular and Learned Culture, 1300–1500*, London: Routledge and Kegan Paul

Kitteredge, G. L. 1958. *Witchcraft in Old and New England*, New York: Russell and Russell

Klaniczay, G. 1990. *The Uses of Supernatural Power*, Oxford: Polity Press

Kluckhohn, C. 1967. *Navaho Witchcraft*, Boston: Beacon Press

Kluckhohn, C. and Leighton, D. 1946. *The Navaho*, Cambridge, Mass.: Harvard University Press

Ladurie, E. Le Roy. 1987a. *Jasmin's Witch*, trans. Brian Pearce, Aldershot: Scolar Press

Bibliography

Ladurie, E. Le Roy. 1987b. *Montailloux. Cathars and Catholics in a French Village, 1294–1324*, trans. Barbara Bray, London: Penguin Books

La Fontaine, J. S. 1994. *The Extent and Nature of Organised and Ritual Abuse. Research Findings*, London: HMSO

Larner, C. 1981. *Enemies of God. The Witch-hunt in Scotland*, Oxford: Basil Blackwell

Larner, C. 1985. *Witchcraft and Religion. The Politics of Popular Belief*, Oxford: Basil Blackwell

Lea, H. C. 1957. *Materials Toward a History of Witchcraft*, Vol. 1, arr. and ed. Arthur C. Howland, New York, London: Thomas Yoseloff

Leach, E. R. 1971. 'Rethinking Anthropology', in E. R. Leach, *Rethinking Anthropology*, University of London: The Athlone Press, pp. 1–27

Lewis, I. M. 1976. *Social Anthropology in Perspective*, Harmondsworth: Penguin Books

Lewis, I. M. 1986. *Religion in Context. Cults and Charisma*, Cambridge: Cambridge University Press

Lienhardt, R. G. 1951. 'Some Notions of Witchcraft among the Dinka', *Africa*, Vol. 21, pp. 303–15

Locke, R. F. n.d. *The Book of the Navajo*, 5th edn, Los Angeles: Mankind Publishing Co.

Luhrmann, T. R. 1989. *Persuasions of the Witch's Craft. Ritual Magic and Witchcraft in Present-day England*, Oxford: Basil Blackwell

Macfarlane, A. 1970a. *Witchcraft in Tudor and Stuart England. A Regional and Comparative Study*, London: Routledge and Kegan Paul

Macfarlane, A. 1970b. 'Witchcraft in Tudor and Stuart Essex', in M. Douglas (ed.), *Witchcraft Confessions and Accusations*, London: Tavistock, pp. 81–99

Macfarlane, A. 1985. 'Foreword' to C. Larner, *Witchcraft and Religion. The Politics of Popular Belief*, Oxford: Basil Blackwell

Maclagan, R. C. 1972. *Evil Eye in the Western Highlands*, Wakefield: E. P. Publishing Ltd

Mair, L. 1969. *Witchcraft*, London: Weidenfeld and Nicolson

Marshall, G. (ed.). 1994. *The Concise Oxford Dictionary of Sociology*, Oxford: Oxford University Press

Marwick, M. G. 1952. 'The Social Context of Ceŵa Witch Beliefs', *Africa*, Vol. xxii, No. 2, pp. 120–35; No. 3, pp. 215–33

Marwick, M. G. 1965. *Sorcery in its Social Setting. A Study of the Northern Rhodesian Ceŵa*, Manchester: Manchester University Press

Marwick, M. G. 1970. 'Witchcraft as a Social Strain Gauge', in M. Marwick (ed.), *Witchcraft and Sorcery*, Harmondsworth: Penguin Books, pp. 280–95

Mayer, P. 1970. 'Witches', in M. Marwick (ed.), *Witchcraft and Sorcery*,

Harmondsworth: Penguin Books, pp. 45–64
Michelet, J. 1960. *Satanism and Witchcraft. A Study in Medieval Superstition* (translation of *La Sorcière*), New York: The Citadel Press
Middleton, J. 1960. *Lugbara Religion: Ritual and Authority among an East African People*, London: Oxford University Press
Middleton, J. 1963. 'Witchcraft and Sorcery in Lugbara', in J. Middleton and E. H. Winter (eds), *Witchcraft and Sorcery in East Africa*, London: Routledge and Kegan Paul, pp. 257–75
Middleton, J. and Winter, E. H. 1963. 'Introduction' to J. Middleton and E. H. Winter (eds), *Witchcraft and Sorcery in East Africa*, London: Routledge and Kegan Paul, pp. 1–26
Midelfort, H. C. E. 1972. *Witch Hunting in Southwestern Germany – The Social and Intellectual Foundations*, Stanford: Stanford University Press
Mitchell, J. C. 1956. *The Yao Village*, Manchester: Manchester University Press
Mitchell, J. C. 1965. 'The Meaning in Misfortune for Urban Africans', in M. Fortes and G. Dieterlen (eds), *African Systems of Thought*, London: Oxford University Press, pp. 192–203
Monter, E. W. 1976. *Witchcraft in France and Switzerland. The Borderlands during the Reformation*, Ithaca and London: Cornell University Press
Morton-Williams, P. 1956. 'The Atinga Cult among the South-Western Yoruba: a Sociological Analysis of a Witch-finding Movement', *Bulletin de l'Ifan*, t. xviii, serie B
Muchembled, R. 1981. *Les Derniers Buchers*, Paris: Ramsay
Murray, M. A. 1921. *The Witch-cult in Western Europe: a Study in Anthropology*, Oxford: Oxford University Press
Murray, M. A. 1934. *The God of the Witches*, London: Faber
Nadel, S. F. 1935. 'Witchcraft and Anti-Witchcraft in Nupe Society', *Africa*, October 1935, pp. 423–47
Nadel, S. F. 1954. *Nupe religion*, London: Routledge & Kegan Paul
Needham, R. 1978. *Primordial Characters*, Charlottesville: University Press of Virginia
Nottenstein, W. 1911. *History of Witchcraft in England from 1558–1718*, Washington: American Historical Association; London: Henry Froude, Oxford University Press
Ochshorn, J. 1994. 'Woman as Witch: the Renaissance and Reformations Revisited', in S. M. Deats and L. T. Lenker, *Gender and Academe. Feminist Pedagogy and Politics*, Lanham: Rowman and Littlefield Publishers, Inc.
Opler, M. E. 1941. *An Apache Life-way. The Economic, Social, and Religious Institutions of the Chiricahua Indians*, Chicago and London:

Bibliography

The University of Chicago Press

Opler, M. E. 1983a. 'The Apachean Culture Pattern and its Origins', in A. Ortiz (ed.), *Handbook of North American Indians, Vol. 10: Southwest*, Washington: Smithsonian Institution, pp. 368–92

Opler, M. E. 1983b. 'Chiricahua Apache', in A. Ortiz, (ed.), *Handbook of North American Indians, Vol. 10: Southwest*, Washington: Smithsonian Institution, pp. 401–18

Ortiz, A. (ed.). 1979. *Handbook of North American Indians, Vol. 9: Southwest*, Washington: Smithsonian Institution

Ortiz, A. (ed.). 1983. *Handbook of North American Indians, Vol. 10: Southwest*, Washington: Smithsonian Institution

Parrinder, G. 1963. *Witchcraft. European and African*, London: Faber

Parsons, T. 1951. *The Social System*, New York: Free Press

Peel, E. and Southern, P. 1969. *The Trials of the Lancashire Witches*, Newton Abbot: David and Charles

Radford, E. and Radford, M. A. 1961. *Encyclopedia of Superstitions*, (ed. and revised by Christina Hole), London: Hutchinson

Randolph, V. 1947. *Ozark Superstition*, New York: Dover Publishing

Rex, J. 1965. *Key Problems of Sociological Theory*, London: Routledge and Kegan Paul

Rex, J. 1987. 'Ethnicity and Race', in P. Worsley (ed.). *The New Introduction to Sociology*, London: Penguin Books, pp. 323–64

Robbins, R. H. 1959. *The Encyclopedia of Witchcraft and Demonology*, New York: Crown Publishers

Rowlands, M. and Warnier, J.-P. 1988. 'Sorcery, Power and the Modern State in Cameroon', *Man (NS)*, Vol. 23, pp. 118–32

Ruel, M. 1970. 'Were-animals and the Introverted Witch', in M. Douglas (ed.), *Witchcraft Confessions and Accusations*, London: Tavistock, pp. 333–50

Russell, J. B. 1972. *Witchcraft in the Middle Ages*, Ithaca and London: Cornell University Press

Russell, J. B. 1981. *A History of Witchcraft: Sorcerers, Heretics, and Pagans*, London: Thames and Hudson

Sabatier, R. 1988. *Blaming Others. Prejudice, Race and Worldwide AIDS*, London: The Panos Institute

Schieffelin, E. L. 1977. *The Sorrow of the Lonely and the Burning of the Dancers*, St. Lucia (Queensland): University of Queensland Press

Scott, L. 1966. *Karnee. A Paiute Narrative*, Reno: University of Nevada Press

Sebald, H. 1978. *Witchcraft. The Heritage of a Heresy*, New York: Elsevier

Service, E. R. 1966. *The Hunters*, Englewood Cliffs: Prentice-Hall Inc.

Seymour, St. J. D. 1992. *Irish Witchcraft and Demonology*, New York: Dorset Press

Simmons, D. 1956. 'An Ethnographic Sketch of the Efik People', in D. Forde (ed.), *Efik Traders of Old Calabar*, London: Oxford University Press, pp. 1–26

Simmons, L. W. (ed.). 1963. *Sun Chief. The Autobiography of a Hopi Indian*, New Haven and London: Yale University Press

Simmons, M. 1980. *Witchcraft in the Southwest. Spanish and Indian Supernaturalism on the Rio Grande*, Lincoln and London: University of Nebraska Press

Solomos, J. 1991. 'Contemporary Forms of Racial Ideology in British Society', *Sage Race Relations Abstracts*, Vol. 16, No. 1, February 1991, pp. 1–15

Soman, A. 1986. 'Trente procès de sorcellerie dans le Perche (1566–1624)', *L'Orne littéraire*, 8, pp. 42–57

Stafford, H. 1953. 'Notes on Scottish Witchcraft Cases, 1590–91', in N. Downs (ed.), *Essays in Honour of Conyers Read*, Chicago: University of Chicago Press, pp. 96–118

Starkey, M. L. 1963. *The Devil in Massachusets: A Modern Inquiry into the Salem Witch Trials*, New York: Trust Books

Steward, J. H. 1938. *Panatubiji' An Owen's Valley Paiute*, Anthropological Papers No. 6, Smithsonian Institution, Bureau of American Ethnology. Washington: United States Government Printing Office

Stewart, C. 1991. *Demons and the Devil. Moral Imagination in Modern Greek Culture*, Princeton: Princeton University Press

Stewart, K. M. 1973. 'Witchcraft among the Mohave Indians', *Ethnology*, Vol. 12, No. 3, July 1973, pp. 315–24

Stewart, K. M. 1983. 'Mohave', in A. Ortiz (ed.), *Handbook of North American Indians, Vol. 10: Southwest*, Washington: Smithsonian Institution, pp. 55–70

Swartz, M. J., Turner, V. W. and Tuden, A. (eds). 1966. *Political Anthropology*, Chicago: Aldine

Terrell, J. U. 1972. *The Navajos. The Past and Present of a Great People*, New York: Perennial Library

Thomas, K. 1970. 'The Relevance of Social Anthropology to the Historical Study of English Witchcraft', in M. Douglas (ed.), *Witchcraft Confessions and Accusations*, London: Tavistock, pp. 47–79

Thomas, K. 1973. *Religion and the Decline of Magic*, London: Penguin Books

Thomas, K. 1975. 'An Anthropology of Religion and Magic, II', *Journal of Interdisciplinary History*, Vol. VI, pp. 91–109

Trevor-Roper, H. R. 1965. *The Rise of Christian Europe*, New York: Harcourt, Brace & World

Trevor-Roper, H. R. 1988. *The European Witch-Craze of the Sixteenth and Seventeenth Centuries*, London: Penguin Books

Truzzi, M. 1972. 'The Occult Revival as Popular Culture: Some Observations on the Old and Nouveau Witch', *Sociological Quarterly*, Vol. 13, Winter, pp. 16–36

Turnbull, C. 1984. *The Forest People*, London: Paladin

Turner, V. W. 1957. *Schism and Continuity in an African Society. A Study of Ndembu Village Life*, Manchester: Manchester University Press

Turner, V. W. 1964. 'Witchcraft and Sorcery: Taxonomy versus Dynamics', *Africa*, Vol. 34, No. 4, pp. 314–25

Upham, C. W. 1867. *Salem Witchcraft*, 2 vols, Boston: Frederick Unger

Vansina, J. 1962. 'A Comparison of African Kingdoms', *Africa*, Vol. 32, pp. 324–35

Wallace, A. F. C. 1972. *The Death and Rebirth of the Seneca*, New York: Vintage Books

Wallace, A. F. C. 1978. 'Origins of the Longhouse Religion', in B. G. Trigger (ed.). *Handbook of North American Indians, Vol. 15: Northeast*, Washington: Smithsonian Institution, pp. 442–8

Weber, M. 1966. *The Theory of Social and Economic Organization* (ed. with an introduction by Talcott Parsons), New York: The Free Press

Whiting, B. B. 1950. *Paiute Sorcery*, Viking Fund Publications in Anthropology, No. 15, New York: Viking Fund

Wilson, M. 1963. *Good Company. A Study of Nyakyusa Age-Villages*, Boston: Beacon Press

Wilson, M. 1970. 'Witch-beliefs and Social Structure', in M. Marwick (ed.), *Witchcraft and Sorcery*, Harmondsworth: Penguin Books, pp. 252–63

Winter, E. H. 1963. 'The Enemy Within: Amba Witchcraft and Sociological Theory', in J. Middleton and E. H. Winter (eds), *Witchcraft and Sorcery in East Africa*, London: Routledge and Kegan Paul, pp. 277–99

Witherspoon, G. 1983. 'Navajo Social Organisation', in A. Ortiz (ed.), *Handbook of North American Indians, Vol. 10: Southwest*, Washington: Smithsonian Institution, pp. 524–35

Index

Advice to Grand Jurymen, 178
African kings
   witchcraft accusations and, 131–2
AIDS, 1, 210–11
Albigensians, 153–4
Alexander III, pope, 158
Alexander IV, pope, 156
ambiguous social positions
   witchcraft accusations and, 94–6
Antera Duke (Efik), 141–3
anti-witch, 20
anti-witchcraft associations, 67
   see also ndakó gboyá
anti-witch movements, 136–7, 201, 203
   see also Atinga cult
anthropology
   witchcraft studies and, 6, 7–8
Apuleius, 145
Aquinas, Thomas, 155, 165
astrologers, 64
Atinga cult, 201–2, 204
Avatip, 54n1
Azande, 48–9, 51, 54, 117, 119, 137
   king, witchcraft against, 53, 131
   misfortune, interpretations of, 74–5
   poison oracle, 54–5, 70–1
   witchcraft and sorcery, 19, 47

Banyang, 111
Baroja, Julio Caro, 192
Basso, Keith H, 79, 87, 115, 118
Baxter, P T W, 25
Beidelman, T O, 41–2, 51
Bemba, 89
   benandante, 187–8
Benedict XI, pope, 157
Boniface VIII, pope, 157, 161–2, 164, 165
Boran, 93, 96
Bothwell, Earl of, 174, 175–6
Briggs, Robin, 8n6, 49n9, 94, 100–1, 194n23
   devins, 61, 62n5, 64, 80
   witch-crazes, 173
   witches, labelling, 120

de Brigue, Jehanne, 166–7
Bryant, A T, 137

Cameroon, 111–12, 205
Canon Episcopi, 148, 167
Carolingian Empire, 148
Cattaneo, Albert, 154
Charlemagne, 148
Charles the Bald, 148
Cartesian philosophy, 189–90
de Charnay, Geoffrai, 162
case studies, 2–4
   interpretation of, 2–4, 17n5, 45n7
Cathars, 153–4, 155, 160
Celts, 146
ceremonial magic, 165
Ceŵa, 84–5
   misfortunes, and witchcraft, 115
   poison oracle, 71, 72
   political competition, 122, 123
   witch suspects, 84, 85–6, 88–9, 102–4, 105, 108–9
   witches, 85
Chattox, 100, 101
   see also Pendle witches
Chilperic III, 148
Christianity
   anti-female bias, 182
Clement V, pope, 157–62
Cohn, Norman, 10, 113, 153, 155, 167n17
Colonna, cardinal Peter, 162
conflict theory, 4–9, 172
   witchcraft and, 92
   see also power
Corrector of Burchard, 148–9
Crawford, J R, 56n4, 96, 109
Crick, Malcolm, 50–1
cross-cultural comparison, 52–3
cunning folk, 61–4, 79–80, 189, 193
cunning men, see cunning folk

Daemonologie, 176, 178
Dark Ages, 146–9
De Praestigiis Daemonum, 189
Demdike, 100, 101

Index

see also Pendle witches
demon worship, 148
Devil, the, 136, 149–50, 155, 175
    see also sabbat
Devil's mark, 33, 72
devil worship
    political accusations of, 153–4, 156, 157–65
diabolic witch (Europe)
    creation of, 169, 171
        power and, 143–4, 168, 200
    definition, 152
    élites and, 169
    extension of accusations and, 181–2
Diana, goddess, 145, 148, 192
    Wild Hunt and, 152
Discoverie of Witchcraft, 176, 189
diviners, 55–6
    accused of witchcraft, 74–83, 181, 188
    diagnoses, 56–64, 69–70
    modern doctors as, 69n9
    Scotland, Western Highlands, 193
    see also cunning folk
doctors, see diviners
Douglas, Mary, 22n7, 26–7, 71–2, 109, 210

Efik, 137, 138–43, 200
elderly,
    witch suspects and, 95–7, 99, 181
England, Early Modern
    familiar spirits, 33
    old, changing attitudes towards, 106–7
    poor, changing attitudes towards, 106–7
    witchcraft, 51
        acquisition of, 48–9
        as antisocial activity, 28–9
        begging and, 88, 108, 192
        Continental beliefs and, 177–8
        curing of, 63–4
        cursing and, 106
        evidence of, 176–7
        executions, 170
        as felony, 176
        local community and, 121
        offenses caused by, 33–4
        quarrels and, 106
        statutes, 172, 177, 190
        trials, 177
        see also Hopkins, Matthew
    witchcraft accusations
        accuser and accused, relative

statuses of, 100–1
        age and, 101
        interpersonal relations, conflict in, 107–8
        marital status and, 101
        neighbours, 106
        women, 99–101
        see also astrologers, cunning folk, Pendle witches
    witch's mark, 33
        searching for, 33, 179
        see also women
England, medieval, 28–9
Europe
    outsiders as witches, 94
Europe, Early Modern
    witch beliefs and social class, 169
    witchcraft
        local community and, 29
        as public offence, 7
    witches
        labelling, 120
        treatment of, 166
    women, and criminal offences, 182
    see also women
Evans-Pritchard, E. E., 19, 28, 47, 54–5, 70
    Azande witchcraft, principles of, 22–3
    witchcraft as mutually reinforcing system of ideas, 116
events,
    different interpretations of, 99

Fabian, superintendant Robert, 18
familiars, 30, 100, 165, 178, 179
    see also England, Early Modern
Favret-Saada, Jeanne, 194, 196, 197, 198
de Floyran, Esquiu, 158–9
Fisiy, Cyprian F, 205
France, Early Modern, 48, 170, 190–1
Franconian Switzerland, 193–4, 196
French-speaking Europe, Early Modern, 34, 80
    see also Briggs, Robin
Fredegond, queen, 146–7
functionalist explanations, 90, 131n1, 213
    criticisms of, 91–2
    witchcraft and, 83–4, 91, 108–9, 130, 203

Ganado Mucho (Navaho leader), 134, 136, 201

Index

Gaul, 146
Gaule, Rev. John, 180
Gbudwe (Azande king), 70–1, 116, 131
Germany, 145–6, 181–2, 184, 186
Ghana, 110–11
Ginzburg, Carlo, 187
Gnostics, 155
Gluckman, Max, 91, 94–5, 101
Good, Sarah, 184
Grandier, Urbain, 65, 72n4
'Great Witch-Craze'
  causes of
    general, 171–2
    local, 173–4, 180
  élites and, 189–90
  end of, 184–5, 189–90
  executions, 170
  property confiscations, 184
  victims, 181–6
Greece, classical, 145
Gregory IX, pope, 154
Gregory of Tours, 146
Guichard, bishop of Troyes, 160–1, 164

Haggard, H Rider, 137
Handsome Lake (Seneca prophet), 135–6
Harris, Rosemary, 71n10, 109n8, 132n3, 141n10
Harrison, Simon, 54n1
Hecate, 145, 152
Heretics of Orleans, 153
Herodias, 152
historians
  witchcraft studies and, 6, 7–8
Hitler, Adolf, 8, 208–9
Holda, 192
Home Circuit, 100n5
Hopi
  doctors/diviners, 59
  elderly, as witches, 96
  witchcraft, tricked into, 48
  witch suspects, 97–99
  witches' society, 45–6
Hopkins, Matthew, 178–80
Horton, Robin, 22n7
house property complex, 95n2
hunter-gatherers, 27

Ibo, 25
image magic, 36
immigrants
  as witch-figures, 1, 209–10
implicit pact, doctrine of, 155, 165

incubus, 161, 163, 165
Innocent III, pope, 153
Innocent VIII, pope, 154, 167
inquisitorial procedure, 166, 169–70
Institoris, Heinreich, 167–8, 171
'introspective' witchcraft, 110
'introverted' witchcraft, 110
Iroquois, 135

James I/VI, 72, 174–6, 178
Jewish world conspiracy
  myth of, 208–9
  as sabbat, 208
Jews
  of Alexandria, 153
  late medieval persecution of, 165
  medieval Europe and, 207–8
  Nazis and, 208–9, 211
  as scapegoats, 8
John XXII, pope, 154, 165
Jones, G I, 25
Jura, 181, 185–6

Kaguru, 41–2, 51, 52
Kaluli, 13–15, 31, 48, 61
Kieckhefer, Richard, 182
Klaniczay, Gábor, 187
Kluckhohn, Clyde, 28, 31, 44, 57
  case material, unreliability of, 80, 96n3
  functional theory of witchcraft, 104
  Navaho
    distrust of extremes, 77
    elderly, position of, 96
    social change and, 201
Knights Templar, 157, 158–62, 164
Konrad of Marburg, 156
kresniks, 187–8
Kyteler, Lady Alice, 162–5

Ladurie, Emmanuel le Roy, 120, 121n6
Larner, Christina
  anthropology and history, 5
  'Great Witch-Craze', 171–2, 190
  local witch-crazes, 173–4
  North Berwick witches, 174
  'sabbath witchcraft' and '*maleficium*', 150n8
  Scottish witch-hunt, 170
  witch suspects, 88, 100
  women, 97
de Ledrede, Richard, 163–4, 208
Lele, 71–2, 119, 136, 201
Lewis, I M, 110–11, 150n7
Lobengula (Ndebele king), 132

Index

Long Parliament, 178–9
Lower Quinton, 17–18
Lowes, John, 179
Lucerne, 10, 113–14
Lugbara, 75
Luhrmann, T R, 8n8, 71n11

McCarthy Senate hearings, 134n5
McCarthy, Senator Joseph, 8
Macfarlane, Alan, x, 28, 33, 193
  church in Middle Ages, 177
  cunning folk, 61, 62, 63, 80
  social change and witchcraft accusations, 108
  witch-suspects, 100
    antisocial behaviour of, 84n7, 88
    begging and, 107, 177
    relation of accuser to accused, 107
    women, 108
Maclagan, Dr R C. 192
Mair, Lucy, x, 48, 69, 72, 117
maleficium, 114, 146, 147–8, 213
  Devil and, 149–50, 165
  diabolic witchcraft and, 150, 165–8, 169
  local community and, 169
  Jura witch trials and, 186
  as 'sorcery', 152
Malleus Maleficarum, 167–8, 171, 181, 200
Mandari, 93, 96
Manichaean heresy, 155
Manuelito (Navaho leader), 134, 136, 201
Map, Walter, 153
marginal social positions and witch-suspects, 96–7
Marwick, M G, 71, 72
  Ceŵa witchcraft cases, 84, 103
  functional theory of witchcraft, 108–9
  witchcraft accusations
    conjugal relationships and, 103
    eccentric behaviour and, 89
    interpersonal conflict and, 105
Mbembe, 132n3
Mbuti, 24
Mein Kampf, 208–9
Michelet, Jules, 186
Middleton, John, ixn1, 95n2
Midelfort, H C Eric, 182, 184–5
midwives
  as witches, 114, 181
minority groups
  as scapegoats, 1–2, 8, 206–7

misfortunes
  conflicting interpretations of, 85–6, 114, 116, 117
  social and personal relationships and, 116
Mitchell, J C, ixn1, 204–5
Mohave, 80–3, 120, 201
de Molay, Jaques, 158–60, 162
Monter, E William, 5, 181n7
Moral Majority, 210
Morton-Williams, P, 202n1, 202n2
Muchembled, R, 121n6, 189n16
Murray, Margaret A, 18, 186–7
Mzilikazi (Ndebele king), 132

Nadel, S F, 68–9, 74, 84n7, 95
Nandi, 93
Navaho, 30–1, 97, 188
  diviners, 57–9
  fee splitting, 77, 89–90
  'frenzy witchcraft', 47
  peyote, 31, 293–4
  social relations, tension in, 104
  suckers, 77–8
  witchcraft
    acquistion of, 48
    cases, 3
    social change and, 201
    as social leveller, 203
  witchcraft accusations
    incidence of, 104
    political use of, 134
  witches
    abnormal behaviour of, 88
    motivations, 47
  witches' society, 42, 44–5
  witch-image, 31
  witch suspects, 77–8, 86–7, 89–90, 93, 96
  'wizardry', 58
ndakó gboyá, 67–9, 74, 137
Ndebele, 137
Ndembu, 122–9. 130
Needham, Rodney, 51–2
Neûchatel witch-craze, 185
de Nogaret, Guillaume, 157–61
Normandy bocage, 194–8
Northern Paiute
  misfortunes, conflicting interpretations of, 114–5
  shamans, 12, 59–61, 69–70
  witch, labelling of, 119
  witchcraft, 12–13
    acquisition of, 48
North Berwick witches, 174–6

– 229 –

Index

Nupe, 67–9, 74, 84n7, 95, 105, 202n2
Nyakyusa, 39–40, 187
  poison ordeal, 71
  repentant witches, 74, 121
  witchcraft, 47
    accusations, 40–1
    inheritance of, 49
    witch-image, 39–41

obsession, 65
  see also Salem Witch-Craze
occult power
  witchcraft suspects and, 114.
  see also cunning folk, diviners
Old Calabar, see Efik
oracles, 54–5, 70–1
ordeals, 71–2, 116, 149
  see also water ordeal
Order of the Solar Temple, 8n7
Osborne, Ruth, 191–2
osculum infame, 151, 153
'outsiders'
  as witch suspects, 28, 183–4

papal inquisition, 154, 156, 172n3
pastoralists, 25, 28, 93, 206
Pendle witches, 34–9, 72, 101
  familiar spirits, 35, 36, 37, 38
  labelling of, 120
  local community and, 121
  Malkin Tower meeting, 37–8
  as wise women, 80
  witchcraft
    aquisition of, 48
    begging and, 107
  witchcraft accusations, 63–4
    quarrels and, 106
personal relations
  witchcraft and conflict in, 101–8
Philip IV, 157–62
le Poer, Arnold, 163–4
Pondo, 39–41
Poor Law, 192
possession, 65–6
power
  conflict and, 6
  Continental with-image and, 152, 168
  defined, 214
  'Great Witch-Craze' and, 171–2
  society and, 5–6
  sources of, 5–6
  witch and, 130–1, 143
  witch beliefs and, 7–9, 53, 168, 200
  witchcraft and, 130–44, 147

witch labelling and, 132, 143, 144
  see also conflict theory, diabolic witch
Proctor, John, 184
prophets
  witchcraft accusations and, 136, 143

Red Jacket (Seneca), 135
ritual magic, 165
Robbins, Rossell Hope, 184
Robert II, 153
Robin (Robert) Artisson (incubus), 163
Roman Law, 156, 173–4
Rome, classical, 145
Rowlands, Michael, 205
de Ruilly, Macète, 166–7
Russell, Jeffrey
  Cathars, 155, 155n11
  Continental witchcraft, 151–2, 154–5
  devil-worship accusations, political use of, 157
  Great Witch-Craze, 170, 171
  'witchcraft' and 'sorcery', 152n10
  witches' organization, existence of, 187, 188

sabbat, 150, 180, 185, 188, 194
  creation of, 154–5
  Devil worship and, 150–1
  Eucharist, parody of, 150–1
  maleficia and, 165
Safwa, 19, 74, 75–7, 187
Salem Witch-Craze, 66–7, 183–4
Salmesbury witches, 38–9
Samburu, 28, 93
'satanists', 8
scapegoats
  selection of, 205, 206, 210, 212
scold, 88n12, 181
Scandinavia, 146
Schieffelin, Edward L, 48
Scotland, 170, 181–2, 190, 191–2
Scot, Reginald, 176, 189
Sebald, Hans, 193
Select Cases of Conscience Touching Witches and Witchcraft, 180
Shaka (Zulu king), 137
shamans, 27, 214
  see also diviners
Shona, 116
  diviners/doctors (nganga), 16, 56–7, 69, 115
  Pentecostal churches, 202–3
  poison ordeal, 71

Index

witch
  motivation of, 47
  self-confessed, 110
  witchcraft, 15–17
  acquisition of, 48
  witchcraft accusations
  elderly and, 96
  political authorities and, 132
  strangers and, 94
  witches' society, 42–4
  witch-image, 30
Slavonic witch beliefs, 145
social change
  witchcraft accusations and, 7, 107–8, 136, 143, 200–5
  see also anti-witch movements
Somali, 28, 93, 96
sorcerer, 18–20
La Sorcière, 186
Soviet Union, 1, 49
Spain, 192
Sprenger, Jacob, 167
Stalin, J V, 8
Stearne, John, 179, 180
strangers/foreigners
  as witch suspects, 93–4, 185, 193
Summis Desiderantes Affectibus, 167
Super Illius Specula, 165
Swanson, G E, 105–6

talion, law of, 149
Tangu
  ranguma, 32, 47, 74, 120
  witch suspects, 86, 88, 93
Tenkswatawa (Shawnee prophet), 136
Teutons, 146
Thirty Years War, 171
Thomas, Keith, x, 28, 193
  church in Middle Ages, 177
  cunning folk, 63
  witchcraft accusations
    antisocial behaviour and, 88
    begging and, 107
    social change and, 108
    witch suspects, 100
    women, predominance of, 108
Tiv, 132–4, 137, 188
trade unions, 210
'traditional' witch, 152
Trevor-Roper, Hugh, 4–5, 171
Truzzi, Marcello, 8n8
Turnbull, Colin, 24
Turner, V W, 124, 126
  Ndembu society, 122
  witchcraft as social catalyst, 109

witches' society, 46
witch-image, 41

Venice, 182–3
Vox in Rama, 156

Waldensians, 154, 156, 160, 188
Walton, Charles, 17–18
Wando (Azande king), 131
Warnier, Jean-Pierre, 205
water ordeal, 149n5, 179, 180
Western Apache, 182
  witchcraft accusations, 104–5, 115, 116, 119–20
  witch suspects, 78–9, 87–8, 90, 93, 96
Weyer, Johan, 189
'white witches', 20
  see also cunning folk
Whiting, Beatrice Blyth, 48
Wilson, Monica, 39–41
Winnemucca (Northern Paiute leader), 13
Winter, E H, 95n2
wise women, see cunning folk
witch
  ambivalence regarding, 41–2
  character of, 1
  Dark Ages and, 146–8
  as deviant, 172
  European, 52
  as false category, 50–1
  local community and, 120–1, 149, 191–2
  motivation of, 10–11, 19–20, 30, 47–8
  self-confessed, 110–12
  as social invention, 114
  sorcerer and, 47
  treatment of, 114, 130–1, 172–3, 191
  see also diabolic witch
witch, detection of
  public opinion and, 64, 69–70
  social relations and, 64, 69–70
  see also anti-witchcraft associations, astrologers, cunning folk, diviners, obsession, oracles, possession
witch, labelling of, 7, 113–29, 118–20, 143, 148, 181, 191, 199–200
  difficulty of, 114–8, 149
  public opinion and, 119
  see also power, witchcraft accusations, witch suspects
witchcraft

– 231 –

Index

acquistion of, 48–9
African, 118
definition, 19–20
as heresy, 7, 28, 151–68
North American, 118
misfortune and, 41, 192
as public offence, 7, 130–1, 136, 144, 168, 200
as social offence, 6–7
sorcery and, 18–19
tricked into, 48
in urban setting, 204–5
see also England, Early Modern; France, Early Modern; French-speaking Europe, Early Modern; 'introspective' witchcraft; 'introverted' witchcraft
witchcraft accusations
accuser and, 105
antisocial behaviour and, 83–90
danger in making accusations, 116
evidence, 115–6, 189
incidence of, 89, 103, 104
non-random nature of, 73–4
pastoralists, 25
political competition and, 110
political use of, 7, 132, 134–6, 138–43, 143–4, 145–7, 174, 200, 205
reconciliation and, 130
structure of society and, 25–8, 41
see also social change, witch suspects
witchcraft beliefs
contemporary Europe and, 192–8
emotional intensity of, 49
misfortune and, 11–12, 18, 41, 49
as self-reinforcing system of ideas, 71, 116
structure of society and, 21, 25–8
see also Europe, Early Modern
witch-craze, 215
see also 'Great Witch-Craze', witch-hunts
'witches' of industrial society (fertility cults, etc.), 8
see also 'satanists'
'witches' of industrial society (scapegoats), 8–9, 205–11, 122

see also AIDS, immigrants, Jews, minority groups
witches' society, 42–6, 111, 186–9
see also sabbat
witch-hunts
indigenous societies and, 134–7
see also 'Great Witch-Craze'
'witch-hunts'
contemporary Europe and, 1–2, 8–9
see also AIDS, immigrants, Jews, minority groups
witch-image, 7, 30–2, 42
application of, 117–18, 143, 144
Continental Europe and, 149–51, 152–3
social structure and, 39–41
symbolic anthropology and, 49–53
see also power, sabbat
witchfinders, see diviners
witch suspects
abnormal behaviour and, 88–90
Africa, 73
categories of, 199
children, 184
cross-cultural variation in, 73–4
Melanesia, 73–4
men, 181–2, 184–5, 186
mutually reinforcing causes, 97
neighbours, 193, 195–6
prominent persons, 184–5
widows, 181
see also ambiguous social positions; diviners; elderly; England, Early Modern; France, Early Modern; French-speaking Europe, Early Modern; midwives; 'outsiders'; strangers/foreigners; witchcraft accusations; women
women
witch suspects and, 94–5, 181, 182, 194
Early Modern England and, 99–101
Early Modern Europe and, 96–7, 99–100

Yeltsin, Boris, 49n10
Yoruba, 201, 202

Zulu, 94–5, 137